A Level of Martin-Lof Randomness

A Level of Martin-Lof Randomness

Bradley S. Tice

Founder and CEO
Advanced Human Design
Cupertino, California
USA

CRC Press
Taylor & Francis Group
Boca Raton   London   New York

CRC Press is an imprint of the
Taylor & Francis Group, an **informa** business

A SCIENCE PUBLISHERS BOOK

CRC Press
Taylor & Francis Group
6000 Broken Sound Parkway NW, Suite 300
Boca Raton, FL 33487-2742

© 2013 Copyright reserved
CRC Press is an imprint of Taylor & Francis Group, an Informa business

No claim to original U.S. Government works

Version Date: 20120327

International Standard Book Number: 978-1-57808-751-8 (Hardback)

Library of Congress Cataloging-in-Publication Data

Tice, Bradley S.
A Level of Martin-Lof randomness/Bradley S. Tice.
   p. cm.
 Includes bibliographical references and index.
 ISBN 978-1-57808-751-8 (hardcover)
1. Kolmogorov complexity. 2. Statistical communication theory.
3. Stochastic processes. I. Title. QA267.7T53 2012
511.3'52--dc23

2011032612

**Visit the Taylor & Francis Web site at
http://www.taylorandfrancis.com
CRC Press Web site at
http://www.crcpress.com**

**Science Publishers Web site at
http://www.scipub.net**

For Mom (proof reader), Lisa (sister)
and Jeff (brother)

Preface

The book will address the notion of compression ratios greater than has been known for random sequential strings in both binary and larger radix based systems as applied to those traditionally found in kolmogorov complexity. The work is the culmination of a decade long research starting with the discovery of a compressible random sequential string in 1998 by the author. The choice of the ternary, quaternary and quinary based systems is a form of functional minimalism that goes back to the reason for a binary level of practical application as found in statistical communication theory and computing. While it is clear the applications of this system to engineering solutions in statistical communication theory and computing are very real, the book will maintain a theoretical statistical level of introduction that is more the realm of physics than engineering.

The introductory chapter to this book explains the reason for the writing of the book as well as its successive chapters. The following chapter will discuss the nature of a Martin-Lof level of randomness and why it is of fundamental importance to statistics. The section on compression addresses the act of compression in a mathematical sense, and as a computational method of reduction of sequential strings. The chapter on radix numbers defines these 'root' numbering systems that will be used to define a random and nonrandom sequential strings. The following chapter will review the notions of symbols and characters as they relate to the structure found in the strings presented in this book.

The next series of chapters address binary; radix 2, ternary; radix 3, quaternary; radix 4, and quinary; radix 5 base numbering systems. Larger radix numbering systems are reviewed in the following section. The summary concludes the presentation of ideas and gives an overview of what was presented in the book. The reference section, also known as the bibliography section, is annotated. I have tried to keep my citations to the minimum as my dissertation (Tice, 2009), and Li and Vitnayi's work (1993/1997), are a wealth of references on this subject. The index is alphabetized and includes important names of people and concepts used and cited in this work.

References

Tice, B.S. (2009) Aspects of Kolmogorov Complexity: The Physics of Information. Gottingen: Cuvillier Verlag.

Li, M. and Vitanyi, P.M.H. (1993/1997) An Introduction to Kolmogorov Complexity and its Applications. New York: Springer.

Contents

List of Images

1

Introduction

The book will address compressible random sequential strings by the use of a simple compression method, or program, that introduces previously unknown levels of compression to random sequential strings in binary and larger radix numbering systems. The presentation will address the material in a purely statistical manner that is more theoretical, physics, rather than applied, engineering, utilizing examples from the binary; radix 2, ternary; radix 3, quaternary; radix 4, and quinary; radix 5, base numbering systems. Larger radix numbers will be addressed in a following chapter. While it is clear that the practical applications of the system to telecommunications, statistical communications theory, and computing are all understood, the nature of this introductory text is theoretical rather than applied.

I have chosen the radix 3, radix 4, and radix 5 levels of base numbers as first; A historical precedent set by the use of the radix 2, or binary, system in Claude E. Shannon's 1948 paper that introduced information theory, also known as statistical communication theory, to the world, and second; a minimalist use of characters is found desirable to the transmission and storage of information, a practical reason, for the use of low radix number bases. I will use larger base numbering systems in a trailing chapter that uses a radix 8, radix 10, radix 12, and radix 16 base number systems in the sequential strings for compression purposes. Most of these radix systems have been tried early in the history of computing.

The summary is the conclusion of the topic and will restate what was presented in a simple and general fashion. The references section is annotated and the index is alphabetized and includes proper names for citation purposes.

The author is the CEO and chief technical researcher at Advanced Human Design that is located in the Central Valley of Northern California, California U.S.A. Dr. Tice has a Doctor of Letters in Teaching (1998) and a Doctor of Philosophy in Physics (2003). Dr. Tice's specialization is in natural languages and telecommunications.

2

Definition of a Level of Martin-Lof Randomness

The following traditional definitions of a non-random and random sequential strings are used in this monograph.

Random

A random sequential string is one that cannot be reduced or compressed from its original sequential length.

Non-random

A non-random sequential string is one that can be reduced or compressed to less that the original sequential string length and be de-compressed to the original sequential string length and retaining the same character type and placement as the non-random original sequential string.

The examples in this work are all finite in nature and keeping with the spirit of A.N. Kolmogorov, a principle foundational researcher in the area of algorithmic randomness, finite objects are more relevant to our experience (Bienvenu, Shafer and Shen, 2009: 27). The definition of randomness will be that of Martin-Lof Randomness as defined by Per Martin-Lof's 1966 paper that defined a new measure for both finite and infinite sequences (Martin-Lof, 1966).

What will be shown in this monograph is a new measure of randomness that is compressible and de-compressible to its original state that is developed from the original work and definitions of randomness as set by Martin-Lof (1966) and Gregory J. Chaitin, who found a common theme with Martin-Lof's notion of randomness in his 1975 paper, and presents both non-random and random sequential strings using various radix based number systems (Bienvenu, Shafer and Shen, 2009: 26 and Chaitin, 1975). This notion of a non-compressible series of random sequential strings and compressible non-random sequential strings, also known as collectives, is most widely known as Kolmogorov Complexity, named after the Russian mathematician A.N. Kolmogorov (Bienvenu, Shafer and Shen, 2009: 14).

3

Compression

Compression is the act of compressing. Decompression is the opposite act of compression.

I have taken the rather arcane nomenclature of a mathematic process, a form of computation, to denote the process of compression in a random or nonrandom sequential string as being a 'compression engine'. This rather Victorian and mechanistic allusion to Babbage's folly is, in essence, a simple computing operation that sorts and stores the original segment, in like groups, and then reduces each subgroup into a notational symbol of the original number of characters found in that specific subgroup (Note #1). Decompression, the act of restoring the original number and type of characters found in the original string.

I have termed the procedure for compressing and de-compressing sequential strings as being a Modified Symbolic Space Program, or MSSP, that was developed in 2006 (Tice, 2009a and 2009b).

The Modified Symbolic Space Multiplier Program

The Modified Symbolic Space Multiplier Program is defined by the following:

1) A space under a character is the character to be multiplied.

2) The amount to be multiplied for the character is defined in the Key Code guide.
3) That character can represent only one value of multiplication although different characters can represent the same value of multiplication through out the string.

Multiplication, the act of repeated addition, has the characters being treated like natural numbers such that the number of a same type character, the multiplicand, is added repeatedly to the number of same type characters to be multiplied, the multiplier, and assigned a value of a single same type character that is underlined to represent the total character quantity for that segment of the string being compressed (Mueller, 1964: 83).

Example

A binary bit string [A] of a character length of 20 bits.

[A] 11000111000001110011
Using the Key Code guide:

Key Code

1 = x3
0 = x3

Three similar characters of a sequential nature, being either all 1's or all 0's, will be assigned an underline to represent the character to be multiplied. The initial character will be underlined.

[A] 11<u>0</u>00<u>1</u>11000<u>0</u>01<u>1</u>10011

In compressing the string, the following characters will be removed from the string:

[A] 11<u>0</u> <u>1</u> 0000<u>0</u>1 0011

The resulting string will contain 14 characters.

[A] 11010000010011

This is a compressed string.

In de-compressing the compressed string, replace the original characters to the compressed string to arrive at the original pre-compressed state of string [A].

[A] 11000111000001110011

The compressed string [A] de-compresses to the original 20 character length with no change to either character placement or character type in string [A].

4

Radix Numbers: A Historical Overview

Radix is a term for root of a number. It is a number or symbol which is made the bases of a scale of numeration (Simpson and Weiner, 1989: 108). Such as in the study of grammar; natural, or human languages, a set of rules defines the root of a word, a morpheme, as being the main or core aspect of the word; morpheme, that is developed by prefixes, added in the initial position of the word, or suffix, added to the end of the word.

The radix base numbers to be examined in this monograph are radix 2, radix 3, radix 4, radix 5, radix 8, radix 10, radix 12, and radix 16. The radix base numbers will be represented by characters that will be non-numeric in a semantic sense and represent only a quality of being opposite of the other characters present in the total radix base number quantity.

A radix, also known as a scale, is a base number that represents a number known as a number system (Weisstein, 2003: 169). The amount of digit symbols used in a system, such as the Arabic numerical system, is termed the radix, or scale, of the system (Richards, 1955: 4). Richards (1955) makes the note that the radix 10 was probably the results of fact the human hands are composed of 10 digits, five digits each hand, and that they are ideal for counting (Richards, 1955: 4). The radix 12, termed a duodecimal system, and has been adopted by clock faces, eggs; a dozen eggs per carton, and the quantity of being gross (Richards, 1955: 4). Richards also notes that the radix numbers two, eight and sixteen afford the best possibilities for an efficient computer (Richards, 1955: 4-5). The radix numbers 2, 3, 8, 10, 12 and 16 are the only radix numbers to have had serious consideration in the design of computer machinery (Richards, 1955: 5). The Standards Committee of the Institute of Radio Engineers has gone to the extravagance of listing the adjectives to be used in describing systems of other radices (Richards, 1955: 5). A digit is a single symbol or character representing an integral quantity and a number is a quantity represented by a group of digits (Richards, 1955: 5)

5

Binary: An Overview of the Radix 2 System

The binary system, also known as a radix 2 based system, is composed of two characters, usually a 0 and a 1, that have no semantic properties except not representing the other. The etymology of the word binary is from the late Latin 'binarius' that was derived from the early Latin 'dvis' (Barnhart, 1988: 94). Group A will represent a non-random sequential binary string and Group B will represent a random sequential binary string. Both Group A and Group B will be 15 characters in total length.

Group A: [000111000111000] (Non-random)

Group B: [001110110011100] (Random)

Utilizing the Modified Symbolic Space Multiplier Program to process like sequential characters, either 0's or 1's, into sub-groups and note them with an underlined character specific to that sub-group and having it represent a specific multiple of that sub-group as found in a key, in this case Group A Key and Group B Key, as a compressed aspect to both Group A and Group B sequential binary strings.

Group A Key: All underlined characters will represent a multiple of 3.

Group B Key: The underlined character 0 will represent a multiple of 2 and the underlined character 1 will represent a multiple of 3.

Group A: [01010]

Group B: [01011010]

The compressed state of Group A, non-random, is five characters in length. The compressed state of Group B, random, is 8 characters in length. Note that the random sequential binary string in Group B compressed to less than the original total pre-compression length. This differs from standards known in Martin-Lof randomness and those found in Kolmogorov Complexity.

Some Examples of Binary Strings

The following are examples of random and non-random sequential strings that can and cannot be compressed using the Modified Symbolic Space Multiplier (MSSM) program.

Example #1

Group C is a random sequential binary string.
Group C: [100111001011101]
15 characters in length.
Group D is a non-random sequential binary string.
Group D: [101010101010101]
15 characters in length.
Group C random is [1 01 0 101 0 1]
Group D non-random is [10]x15

Example #2

Group E is a random sequential binary string.
Group E: [001011000111010]

15 characters in length.
Group F is a non-random sequential binary string.
Group F: [000111000111000]
15 characters in length.
Group E is random [0 101 0 1 0 1 0]
Group F is non-random [01010]

Example #3

Group G is a random sequential binary string.
Group G: [11101100100011010110]
20 characters in length.
Group H is a non-random sequential binary string.
Group H: [11001100110011001100]
20 characters in length.
Group G is random [1 01 01 0 1 1 1 0]
Group H is non-random [1010101010]

Example #4

Group I is a random sequential binary string.
Group I: [11110111001100010000]
20 characters in length.
Group J is a non-random sequential binary string.
Group J: [00001111000011110000]
20 characters in length.
Group I is random [1 01 1 10 10]
Group J is non-random [01010]

Example #5

Group K is a random sequential binary string.
Group K: [00001101110010100011]
20 characters in length.

Group L is a non-random sequential binary string.
Group L: [11111000001111100000]
20 characters in length.
Group K is random [010 1 0 101 0 1]
Group L is non-random [1010]

Example #6

Group M is a non-random sequential binary string.
Group M: [1010101010]
10 characters in length.
Group N is a random sequential binary string.
Group N: [1000110011]
10 characters in length.
Group M is non-random [10]x5.
Group N is random [10 1 0 1]

The binary sequential string can have the dual properties of being both random and non-random properties such as in Group A below:

Group A: [101010101110011]

15 character in length

The first 8 characters are non-random and the second set of characters are random. This mixing of both random and non-random sets of characters produces these qualities in a binary sequential string. This is for both indefinite, infinite, and finite binary sequential strings.

The mixing of both non-random and random binary sequential strings are typical for long chains of binary sequential strings that are processed by computers.

The use of both non-random and random binary sequ strings can be seen in the following groups:

Group #B: [101010111000]

Note Group #B can be considered both non-random in the first part and non-random and random in the second part of the sequence.

Group #C: [111011000110111000111000]

Again the segment is divided into random and non-random binary sequential segments.

Group #D: [0000000000000]

A common non-random binary sequence.

Group #E: [000000000000000000000001]

Even the slight addition of a single differing character makes this binary sequential string random.

6

Ternary: An Overview of the Radix 3 System

The ternary, or radix 3 base system, is also known as a trit, a development from the radix 2, binary, nomenclature.

A ternary, or radix 3, based system is defined as three separate characters, or symbols, that have no semantic meaning apart from not representing the other characters. This is the same notion Shannon gave to the binary based system used in his paper's on information theory upon it's publication in 1948 (Shannon, 1948). Richards has noted that the radix 3 based system as the most efficient base, more so than even the radix 2 or radix 4 based systems (Richards, 1955: 8-9).

A ternary, or radix 3, based system there are three characters used that have no semantic meaning except not representing the other two characters. Group C will represent a non-random ternary sequential string and Group D will represent a random ternary sequential string. The total length for each group, Group C and Group D, will be 12 characters in length. The three characters to be used in this study are a 0, 1, and 2.

Group C: [001122001122]

Group D: [001222011222]

Again each group will be assigned a specific compression multiple based on a specific character type, in this case an underlined 0, 1, and a 2, as defined in a key, Group C Key and Group D Key.

Group C Key: The underlined characters 0, 1 and 2 will represent each a multiple of 2.

Group D Key: The underlined character 0 will represent a multiple of 2. The underlined character 1 will represent a multiple of 2 and the underlined character 2 will represent a multiple of 3.

Group C: [012012]

Group D: [012012]

The compressed state of Group C, non-random, is 6 characters in length. The compressed string of Group D, random, is 6 characters in length. Again note that Group D, the random sequential ternary string, is less than it's pre-compressed state, and again, is novel for those extrapolations of binary examples found in Kolmogorov Complexity.

Some Examples of Ternary Sequential Strings

The following are some examples of random and non-random ternary sequential strings that can and cannot be compressed using the Modified Symbolic Space Multiplier (MSSM) program.

Example #1

Group E is non-random and Group F is random.

Group E Key: Characters, 0, 1 and 2 will all have the value of a multiple of three if underlined.

Group F Key: The character 1 will have a value of a multiple of two if underlined and the character 2 will have a value of a multiple of three if underlined.

Group E: [000111222000111222]

Group F: [011122200111222]

The Compressed State of Group E and Group F.

Group E: [012012]

Group F: [0120012]

The resulting compressed state of Group E is 6 characters in length from the original 18 character length. The compressed state of Group F is 7 characters in length from an original character length of 15.
Note that the compressed state of the random Group F is less than half of the original length of 15 characters.

7

Quaternary: An Overview of the Radix 4 System

The quaternary system is also known as the Radix 4 base number system. The following examples are random and non-random compressed and non-compressed quaternary sequential strings.

Non-compressed quaternary non-random sequential string.

Example A: [111000222111000222]

Compressed Non-compressed non-random quaternary sequential string.

Example A: [1 0 2 1 0 2]

Non-compressed quaternary random sequential string.

Example B: [1111002222211]

Compressed quaternary random sequential string.

Example B: [1 0 2 1]

Non-compressed quaternary random and non-random sequential string.

Example C: [111000222111000022222]

Compressed quaternary random sequential string.

Example C: [1 0 2 1 0]

As with the other radix based number systems, the quaternary system can be used on both non-random and random sequential strings. The distribution is higher in the infinite sequential string rather than the finite string.

The following are some more examples of both random and non-random sequential strings.

Non-compression quaternary non-random sequential string.

Example D: [22220000111122220000]

Compressed quaternary non-random sequential string.

Example D: [2 0 1 2 0]

Non-compressed quaternary random sequential string.

Example E: [11102222200011111]

Compressed quaternary random sequential string.

Example E: [1 02 0 1]

Quinary: An Overview of the Radix 5 System

A radix 5 base is composed of five separate symbols with each an individual character with no semantic meaning. A random string of symbols has the quality of being 'less patterned' than a non-random string of symbols. Traditional literature on the subject of compression, the ability for a string to reduce in size while retaining 'information' about its original character size, states that a non-random string of characters will be able to compress, were as the random string of characters will not compress.

The following examples will use the following symbols for a radix 5 based system of characters [Example A].
Example A
o
O
Q
1
I

The following is an example of compression of a random and non-random radix 5 base system. A non-random string of radix 5 based characters with a total 15 character length [Group A].

Group A oooOOOQQQ111III

A random string of a radix 5 based characters with a total of 15 character length [Group B].

Group B oooOOQQQQ11IIIIII

If a compression program were to be used on Group A and Group B that consisted of underlining the first individual character of a similar group of sequential characters, moving towards the right, on the string and multiplying it by a formalized system of arithmetic as found in a Key, see Key Code A and key Code B, with the compression of Group A and Group B as the final result.

Key Code A (For Group A)

o = x 3
O = x 3
Q = x 3
1 = x 3
I = x 3

Group AoOQ1I

Resulting in a 5 character length for Group A.

Key Code (For Group B)

o = x 3
O = x 2
Q = x 4
1 = x 2
I = x 4

Group B oOQ1I

Resulting in a 5 character length for Group B.

Both Group A (Non-random) and Group B (Random) have the same compression values, each group resulted in a compression value of 1/3 the total pre-compression, original, state. This contrasts traditional notions of random and non-random strings.

These findings are similar to Tice (2003) and have applications to both Algorithmic Information Theory and Information Theory.

Some other examples using Example A Radix 5 characters [oOQ1I] to test random and non-random sequential strings.

The following is a non-random string of a radix 5 based characters with a total of 15 character length [Group A].

Group A oooOOOQQQ111III

A random string of a radix 5 based characters with a total of 15 character length (Group C).

Group C oooooOQQQQQ1I

If a compression program were to be used on Group A and Group C that consist of underlining the first individual character of a similar group of sequential characters, moving towards the right, on the string and multiplying it by a formalized system of arithmetic as found in a key, see Key Code A and Key Code C, with the compression of Group A and Group C as the final result.

Key Code C (For Group A)

o = x 3
O = x 3
Q = x 3
1 = x 3
I = x 3

Group A o̲O̲Q̲1̲I̲

Resulting in a 5 character length for Group A.

Key Code C (For Group C)

o = x 5
O = x 1

```
Q = x 5
1 = x 1
I = x 1
```

Group C o̲OQ1I

Resulting in a 5 character length for Group C.

This example has Group A as a non-random string and Group D as a random string using radix 5 characters for a total 15 character length.

A non-random string of radix 5 characters with a 15 character length (Group A).

Group A oooOOOQQQ111III

A random string of a radix 5 based characters with a total of 15 character length (Group D).

Group D oOOOOQQ1111IIII

If a compression program were to be used on Group A and Group D that consisted of underlining the first individual character of a similar group of sequential characters, moving towards the right, on the string and multiplying it by a formalized system of arithmetic as found in a key, see Key Code A and Key Code D, with the compression of Group A and Group D as the final result.

Key Code A (For Group A)

```
o = x 3
O = x 3
Q = x 3
1 = x 3
I = x 3
```

Group A o̲OQ1I

Resulting in a 5 character length for Group A.

Key Code D (For Group D)

o = x 1
O = x 4
Q = x 2
1 = x 4
I = x 4

Group D o̲O̲Q̲1̲I̲

Resulting in a 5 character length for Group D.

As a final example Group A is a non-random sequential string and Group E as a random sequential string using a radix 5 characters for a total of 15 character length.

A non-random string of radix 5 based characters with a 15 character length (Group A).

Group A oooOOOQQQ111III

A random string of a radix 5 based characters with a total of 15 character length (Group E).

Group E ooOOOOQQQQ111II

If a compression program were to be used on Group A and Group E that consisted of underlining the first individual character of a similar group of sequential characters, moving to the right, on the string and multiplying it by a formalized system of arithmetic as found in a key, see Key Code A and Key Code E, with the compression of Group A and Group E as the final result.

Key Code A (For Group A)

o = x 3
O = x 3
Q = x 3
1 = x 3
I = x 3

Group A <u>oOQ1I</u>

Resulting in a 5 character length for Group A.

Key Code E (For Group E).

o = x 2
O = x 4
Q = x 4
1 = x 3
I = x 2

Group E <u>oOQ1I</u>

Resulting in a 5 character length for Group E.

Again, these examples conflict with traditional notions of random and non-random sequential strings in that the compression ratio is one third that of the original character number length for both the random and non-random sequential strings using a radix 5 base system.

9

Larger Radix Numbers

The larger radix numbers are those number systems that are beyond the radix 5 level and will, for the purposes of this chapter, start with the radix 8 and proceed to the radix 10, radix 12, and the radix 16 based numbering systems. In many respects this section of the book takes on large redundant number chains that while appear unnecessary for a book, and reflect a more practical application in a digital format, i.e., on a computer, than can be reflected in manuscript form.

Nonetheless, these large radix numbers will be utilized with the compression program to obtain compressible random sequential strings for the purpose of examination. I have used the work of Richards (1955) and Knuth (1998) as a guide to the importance of these base numbering systems. Richards (1955) has noted that the radix's 2, 3, 8, 10, 12, and 16 are the only radix numbers to be considered for computing machinery (Richards, 1955: 5).

I will use the radix 8, radix 10, radix 12 and radix 16 for character types in this section keeping with the historical nature of use of these radix systems in mechanical instrumentation and calculating machines.

The Radix 8 Base Number System

Knuth cites Charles XII of Sweden as to consider the use of the radix 8 system for calculations in 1717, but died in a battle before decreeing such a system (Knuth, 1998: 200). The radix 8 is known as an octal system from the early English octonary or octonal (Knuth, 1998: 201).

The Modified Symbolic Space Multiplier Program will be used on both random and non-random sequential strings using the radix 8 based system.

The radix 8 based system will use the following characters: [0, 1, 2, 3, 4, 5, 6, and 7].

A non-random radix 8 sequential string is represented in Example [A].

Example [A]:

[0000000011111111222222223333333344444444555555556666666677777777]

A random radix 8 sequential string is represented in Example [B].

Example [B]:

[00001112222222233334444444455556666667777]

Using Example [A] of a non-random sequential string the following compression can be done using the Key Code guide.

Example [A]:

[0000000011111111222222223333333344444444555555556666666677777777]

Key Code Guide

```
0 = x 8
1 = x 8
2 = x 8
3 = x 8
4 = x 8
5 = x 8
6 = x 8
7 = x 8
```

With the resulting compression of non-random radix 8 Example [A]: [01234567]

Using the Modified Symbolic Space Multiplier Program on a random radix 8 sequential string as found in Example [B]:

Example [B]:

[00001112222222233344444444455556666667777]

Using the following Key Code Guide for the compression of random radix 8 Example [B]:

Key Code Guide

```
0 = x 4
1 = x 3
2 = x 8
3 = x 3
4 = x 8
5 = x 4
6 = x 6
7 = x 4
```

With the following compressed state of the random radix 8 Example [B]:

[01234567]

De-compression will result in the original state of both the non-random and random radix 8 sequential strings.

The Radix 10 Base Number System

The radix 10 is termed a denary or decimal system (Knuth, 1998: 201). The radix 10 is a ten character system.

The Modified Symbolic Space Multiplier Program will be used on both random and non-random radix 10 based sequential strings. Example [C] will represent the non-random radix 10 character sequential string.

Example [C]: [00001111222233334444555566667777788 889999]

Example [D] will represent the random radix 10 character sequential string.

Example [D]: [001112222333334444445555555566666666 777777777888899]

The non-random radix 10 sequential string will use the following Key Code Guide:

Key Code Guide

```
0 = x 4
1 = x 4
2 = x 4
3 = x 4
4 = x 4
5 = x 4
6 = x 4
7 = x 4
8 = x 4
9 = x 4
```

Resulting in the following compressed state of Example [C]:

Example [C]: [0123456789]

The following Key Code Guide will be used for random radix 10 sequential string as found in Example [D].

Key Code Guide

0 = x 2
1 = x 3
2 = x 4
3 = x 5
4 = x 6
5 = x 7
6 = x 8
7 = x 9
8 = x 4
9 = x 2

With a compressed radix 10 base as found in Example [D]:

Example [D]: [0123456789]

The Radix 12 Base Number System

Richards (1955) notes that the radix 12 has the inherent property of being divisible by more numbers than any other small integer (Richards, 1955: 4).

The Modified Symbolic Space Multiplier Program will be used on both random and non-random radix 12 based sequential strings.

Example [E] will represent the non-random radix 12 character sequential string.

Example [E]: {000111222333444555666777888999AAABBB}

Example [F] will represent the random radix 12 character sequential string.

Example [F]: {0011112223445555666777788999ABB}

The non-random radix 12 sequential string will use the following Key Code Guide:

Key Code Guide

0 = 3
1 = 3
2 = 3
3 = 3
4 = 3
5 = 3
6 = 3
7 = 3
8 = 3
9 = 3
A = 3
B = 3

Resulting in the following compressed state of Example [E]:

Example [E]: {0123456789AB}

The following Key Code Guide will be used for a random radix 12 sequential string as found in Example [F].

Key Code Guide

0 = 2
1 = 4
2 = 3
3 = 1
4 = 2
5 = 5
6 = 3
7 = 4
8 = 2
9 = 3
A = 1
B = 2

With a compressed radix 12 base as found in Example [F]:

Example [F}: {<u>0123456789AB</u>}

The Radix 16 Base Number System

Perhaps the most eccentric historical radix number is the Radix 16 that has the Swedish-American civil engineer John W. Nystrom devising a radix 16 base for numeration, weights, and measures that also had it's own pronunciation system, called a 'Tonal System' in 1863 (Knuth, 1998: 201).

The Modified Symbolic Space Multiplier Program will be used on both random and non-random radix 16 based sequential strings. **Example [G]** will represent the non-random radix 16 character sequential string.

Example [G]: {00112233445566778899AABBCCDDEEFF}

Example [H] will represent the random radix 16 character sequential string.

Example [H]:
{000122333344444455666788899999ABBCCCCDDEEEEEFF}

The non-random radix 16 sequential string will use the following Key Code Guide:

Key Code Guide

0 = 2
1 = 2
2 = 2
3 = 2
4 = 2
5 = 2
6 = 2
7 = 2
8 = 2
9 = 2

```
A = 2
B = 2
C = 2
D = 2
E = 2
F = 2
```

   Resulting in the following compressed state of Example [G]:

Example [G]: {0123456789ABCDEF}

The following Key Code Guide will be used for a random sequential radix 16 sequential string as found in Example [H].

Key Code Guide

```
0 = 3
1 = 1
2 = 2
3 = 4
4 = 5
5 = 2
6 = 3
7 = 1
8 = 3
9 = 5
A = 1
B = 2
C = 4
D = 2
E = 5
F = 2
```

   Resulting in the following compressed state of Example [H]:

Example [H]: {0123456789ABCDEF}

10

Universal and Truncated Applications

Many of the examples of compressed sequential strings in this monograph have arbitrarily used both all characters in a sequential string for compression resulting in a maximal compression of the original length of that sequential string giving 'universal' compression to the string or only some of the characters in a compressed sequential string, sub-groups within the total string length, that are only part of the whole string, a truncation of the sequential string, into a semi-compression of the total sequential string.

The reason for this is flexibility of use of the compression algorithm. Not all real world applications of a compression algorithm involve all characters in a sequential string. Examples from the fields of engineering and genetics can be found in this monograph in section D and E of the appendix.

While this monograph was written as an abstract mathematical and statistical text only, the examples given in the above referenced section have practical everyday applications that can further the engineering and natural science disciplines. Past work in using this compression algorithm has been in the field of telecommunications and statistical communication theory in particular and I have tried to stay clear of the 'rude mechanics' of applications in either the engineering or science fields that would be better placed in a textbook targeted for that specific audience. For now, this monograph will be abstract and theoretical in nature giving to the foundational needs of the material for such a fundamental statistical function.

The very nature of this compression algorithm is very robust and flexibly in that it can be applied to both random and non-random sequential strings as well as compression of those complete string of characters or partial compression of some characters as the need arises in both random and non-random strings.

11

Conclusions

Traditional literature on algorithmic randomness has defined a random sequential string as being non-reducible, not compressible, and that only a non-random sequential string is compressible and de-compressible to its original state (See Note #2). This monograph has raised salient points in reconsidering the rather rigid definition of randomness to include those algorithmic systems that have random compressible sequential strings.

Perhaps, as taking some of my ideas from my dissertation work on language and Godel's theorem, the inclusion of the word 'may' be compressible, rather than the all pervasive 'will' be compressible to a random property to a definition of a random sequential string (Tice, 2008). The following new definition could look something like the following:

Random

A random sequential string may be compressible based on the type of algorithmic system used as a program for that random sequential string.

While this is only a suggestion, the nature of this suggestion is of fundamental importance to the very basic question of what is the definition of randomness as found in a sequential string? The work found in this monograph has shown that the traditional notion of a random sequential string as being invalid and that the measures of randomness found in the work of the author to be more precise and specific as in defining the nature of a random sequential string.

12

Summary

The text has presented a model of randomness that is lower than the traditional norms found to measure a level of Martin-Lof randomness in a sequential string. This marks a new standard for the measure of randomness in a sequential string. The examples give a clear representation of the nature of this randomness as found in the compression and decompression of various radix sequential strings.

References

Barnhart, R.K. (1988) The Barnhart Dictionary of Etymology. New York: H.W. Wilson Company.

Bienvenu, L., Shafer, G. and Shen, A. (2009) "On the history of martingales in the study of randomness". Electronic Journal for the History of Probability and Statistics, Volume 5, Number 1, June 2009, pp. 1-40. A well written account of the early history of algorithmic randomness.

Chaitin, G.J. (1975) "A theory on program size formally identical to information theory". Journal of the ACM, Volume 22, pp. 329-340. Chaitin's 1975 paper parallels Martin-Lof's paper of 1966.

Knuth, D.E. (1998) The Art of Computer Science: Volume 2: Semi-numerical Algorithms. Menlo Park, California: Addison-Wesley. An excellent reference, although a little dated in some areas.

Kotz, S. and Johnson, N.I. (1982) Encyclopedia of Statistical Sciences. New York: John Wiley & Sons, Inc. A good source for encyclopedia level searches on statistics.

Li, M. and Vitanyi, P.M.H. (1993/1997) An Introduction to Kolmogorov Complexity and its Applications. New York: Springer. The main source for information on Kolmogorov Complexity. Updated edition to be publishing in 2009.

Martin-Lof, P. (1966) "The definition of random sequences". Information and Control, Volume 9, pp. 602-619. Defining paper on a definition of a random sequence. Chaitin (1975) also arrived at the same conclusion.

Mueller, F.J. (1964) Arithmetic: Its Structure and Concepts. Englewood Cliffs, New Jersey: Prentice Hall.

Richards, R.K. (1955) Arithmetic Operations in Digital Computers. Princeton: D. Van Nostrand Company, Inc. I found this to be well written and a wealth of difficult to find information from the early years of computing.

Seibt, P. (2006) Algorithmic Information Theory: Mathematics of Digital Information Processing. New York: Springer.

Shannon, C.E. (1948) "A mathematical theory of communication". Bell Technical Journal, Volume 27, pp. 379–423 & pp. 623–656. Seminal papers on what started information theory. Published as a book, with Warren Weaver, in 1949. Still in print.

Simpson, J.A. and Weiner, S.C. (1989) The Oxford English Dictionary. Oxford: Clarendon Press.

Tice, B.S. (2003) Two Models of information. Bloomington: 1st Books Library. The author's first published work on algorithmic complexity.

Tice, B.S. (2008) Language and Godel's Theorem. Maastricht, The Netherlands: Shaker Verlag. The author's published mathematics dissertation.

Tice, B.S. (2009a) Aspects of Kolmogorov Complexity: The Physics of Information. Gottingen: Cuvillier Verlag. The authors dissertation on the subject.

Tice, B.S. (2009b) Aspects of Kolmogorov Complexity: The Physics of information. Denmark: River Publishers. A revised edition of the authors dissertation.

Weisstein, E.W. (2003) CRC Concise Encyclopedia of Mathematics. New York: chapman & Hall/CRC.

Notes

Note #1

Charles Babbage designed, but did not construct, a calculating machine he called his 'Analytical Engine' in the early part of the 19th Century in England (Knuth, 1998: 201). Currently a United Kingdom campaign is underway to build a prototype of Charles Babbage's Analytical Engine (Fildes, J. (2010) "Campaign builds to construct Babbage Analytical Engine" BBC News Online, October 14, 2010, Website: http:///www.bbc.co.uk/news/technology-11530905?print=true, pp. 1-4).

Note#2

The compression algorithm found in this monograph is the most precise and accurate level of randomness of a sequential string known in the field of statistics. I made a world record claim to Guinness World Records for this distinction on February 12, 2010.

Appendices

Compression and Geometric Data*

*A poster given at the 2009 SIAM/ACM Joint Conference on Geometric and Physical Modeling, Hilton Hotel Financial District, San Francisco, California U.S.A. October 5-8, 2009. Poster session PPO Monday October 5, 2009 "Compression and Geometric Data" author Bradley S. Tice.

Compression and Geometric Data

Bradley S. Tice

ABSTRACT

Kolmogorov Complexity defines a random binary sequential string as being less patterned than a non-random binary sequential string. Accordingly, the non-random binary sequential string will retain the information about it's original length when compressed, where as the random binary sequential string will not retain such information. In introducing a radix 2 based system to a sequential string of both random and non-random series of strings using a radix 2, or binary, based system. When a program is introduced to both random and non-random radix 2 based sequential strings that notes each similar subgroup of the sequential string as being a multiple of that specific character and affords a memory to that unit of information during compression, a sub-maximal measure of Kolmogorov Complexity results in the random radix 2 based sequential string. This differs from conventional knowledge of the random binary sequential string compression values.

PACS numbers: 89.70 Eg, 89.70 Hj, 89.75 Fb, 89.75 Kd

Traditional literature regarding compression values of a random binary sequential string have an equal measure to length that is not reducible from the original state [1]. Kolmogorov complexity states that a random sequential string is less patterned than a non-random sequential string and that information about the original length of the non-random string will be retained after compression [2]. Kolmogorov complexity is the result of the development of Algorithmic Information Theory that was discovered in the mid-1960's [3]. Algorithmic Information Theory is a sub-group of Information Theory that was developed by Shannon in 1948 [4].

Recent work by the author has introduced a radix 2 based system, or a binary system, to both random and non-random sequential strings [5]. A patterned system of segments in a binary sequential string as represented by a series of 1's and 0's is rather a question of perception of subgroups within the string, rather than an innate quality of the string itself. While Algorithmic Information Theory has given a definition of patterned verses patternless in sequential strings as a measure of random verses non-random traits, the existing standard for this measure for Kolmogorov Complexity has some limits that can be redefined to form a new sub-maximal measure of Kolmogorov Complexity in sequential binary strings [6]. Traditional literature has a non-random binary sequential string as being such: [111000111000111] resulting in total character length of 15 with groups of 1's and 0's that are sub-grouped in units of threes. A random binary sequence of strings will look similar to this example: [110100111000010] resulting in a mixture of sub-groups that seem 'less patterned' than the non-random sample previously given.

Compression is the quality of a string to reduce from it's original length to a compressed value that still has the property of 'decompressing' to it's original size without the lose of the information inherent in the original state before compression. This original information is the quantity of the strings original length before compression, bit length, as measured by the exact duplication of the 1's and 0's found in that original sequential string. The measure of the string's randomness is just a measure of the patterned quality found in the string.

The quality of 'memory' of the original pre-compressed state of the binary sequential string has to do with the quantity of the number of 1's and 0's in that string and the exact order of those digits in the original string are the measure of the ability to compress in the first place. Traditional literature has a non-random binary sequential string as being able to compress, while a random binary sequential string will not be able to compress. But if the measure of the number and order of digits in a binary sequence of strings is the sole factor for defining a random or non-random trait to a binary sequential string, then it is possible to 'reduce' a random binary sequential string by some measure of itself in the form of sub-groups.

These sub-groups, while not being as uniform as a non-random sub-group of a binary sequential string, will nonetheless compress from the original state to one that has reduced the redundancy in the string by implementing a compression in each sub-group of the random binary sequential string. In other words, each sub-group of the random binary sequential string will compress, retain the memory of that pre-compression state, and then, When decompressed, produce the original number and order to random binary sequential string.

The memory aspect to the random binary sequential string is, in effect, the retaining of the number and order of the information found in the original pre-compression state. This can be done by assigning a relation to the subgroup that has a quality of reducing and then returning to the original state that can be done with the use of simple arithmetic. By assigning each sub-group in the random binary sequential string with a value of the multiplication of the amount found in that sub-group, a quantity is given that can be retained for use in reducing and expanding to the original size of that quantity and can be represented by a single character that represents the total number of characters found in that sub-group.

This is the very nature of compression and duplicates the process found in the non-random binary sequential strings. As an example the random binary sequential string [110001001101111] can be grouped into sub-groups as follows: {11}, {000}, {1}, {00}, {11}, {0}, and {111} with each sub-group bracketed into common families of like digits. An expedient method to reduce this string would be to take similar types and reduce to a single character that represented a multiple of the exact number of characters found in that sub-group. In this case taking the bracketed {11} and assign a multiple of 2 to a single character, 1, and then reduced it to a single character in the bracket that is underlined to note the placement of the compression. The compressed random binary sequential string would appear like this: [1000100101111] with the total character length of 13, exhibiting the loss of two characters due to the compression of the two similar sub-groups.

De-compression would be the removal of the underlining of each character and the replacement of the 1's characters to each of the sub-groups that would constitute a 100% retention of the original character number and order to the random binary sequential string. This makes for a new measure of Kolmogorov Complexity in a random binary sequential string.

Summary

The use of a viable compression method for sequential binary strings has applied aspects to transmission and storage of geometric data. Future papers will explore practical applications to industry regarding applied aspects of compression to geometric data.

Reference

[1] Kotz, S. and Johnson, N.I., Encyclopedia of Statistical Sciences (John Wiley & Sons, New York, 1982).

[2] abide.

[3] Solomonoff, R.J., Inf. & Cont. 7, 1-22 & 224-254 (1964), A.N. Kolmogorov, Pro. Inf. & Trans. 1, 1-7 (1965) and G.J. Chaitin, Jour. ACM 16, 145-159 (1969).

[4] Shannon, C.E., Bell Labs. Tech. Jour. 27, 379-423 and 623-656 (1948).

[5] Tice, B.S., Aspects of Kolmogorov Complexity: The Physics of Information. (River Publishers, Denmark, 2009).

[6] Kotz, S. and Johnson, N.I., Encyclopedia of Statistical Sciences (John Wiley & Sons, New York, 1982).

Random and Non-random Sequential Strings Using a Radix 5 Base System

Bradley S. Tice

Kolmogorov Complexity defines a random binary sequential string as being less patterned than a non-random binary sequential string. Accordingly, the non-random binary sequential string will retain the information about it's original length when compressed, where as the random binary sequential string will not retain such information. In introducing a radix 5 based system to a sequential string of both random and non-random series of strings using a radix 5, or quinary, based system. When a program is introduced to both random and non-random radix 5 based sequential strings that notes each similar subgroup of the sequential string as being a multiple of that specific character and affords a memory to that unit of information during compression, a sub-maximal measure of Kolmogorov Complexity results in the random radix 5 based sequential string. This differs from conventional knowledge of the random binary sequential string compression values.

PACS numbers: 89.70Eg, 89.70Hj, 89.75Fb, 89.75Kd

Traditional literature regarding compression values of a random binary sequential string have an equal measure to length that is not reducible from the original state [1]. Kolmogorov complexity states that a random sequential string is less patterned than a non-random sequential string and that information about the original length of the non-random string will be retained after compression [2]. Kolmogorov complexity is the result of the development of Algorithmic Information Theory that was discovered in the mid-1960's [3]. Algorithmic Information Theory is a sub-group of Information Theory that was developed by Shannon in 1948 [4].

Recent work by the author has introduced a radix 5 based system, or a quinary system, to both random and non-random sequential strings [5]. A patterned system of segments in a binary sequential string as represented by a series of 1's and 0's is rather a question of perception of subgroups within the string, rather than an innate quality of the string itself. While Algorithmic Information Theory has given a definition of patterned verses patternless in sequential strings as a measure of random verses non-random traits, the existing standard for this measure for Kolmogorov Complexity has some limits that can be redefined to form a new sub-maximal measure of Kolmogorov Complexity in sequential binary strings [6]. Traditional literature has a non-random binary sequential string as being such: [111000111000111] resulting in total character length of 15 with groups of 1's and 0's that are sub-grouped in units of threes. A random binary sequence of strings will look similar to this example: [110100111000010] resulting in a mixture of sub-groups that seem 'less patterned' than the non-random sample previously given.

Compression is the quality of a string to reduce from it's original length to a compressed value that still has the property of 'decompressing' to it's original size without the loss of the information inherent in the original state before compression. This original information is the quantity of the strings original length before compression, bit length, as measured by the exact duplication of the 1's and 0's found in that original sequential string. The measure of the string's randomness is just a measure of the patterned quality found in the string.

The quality of 'memory' of the original pre-compressed state of the binary sequential string has to do with the quantity of the number of 1's and 0's in that string and the exact order of those digits in the original string are the measure of the ability to compress in the first place. Traditional literature has a non-random binary sequential string as being able to compress, while a random binary sequential string will not be able to compress. But if the measure of the number and order of digits in a binary sequence of strings is the sole factor for defining a random or non-random trait to a binary sequential string, then it is possible to 'reduce' a random binary sequential string by some measure of itself in the form of sub-groups. These sub-groups, while not being as uniform as a non-random sub-group of a binary sequential string, will nonetheless compress from the original state to one that has reduced the redundancy in the string by implementing a compression in each subgroup of the random binary sequential string. In other words, each sub-group of the random binary sequential string will compress, retain the memory of that pre-compression state, and then, when decompressed, produce the original number and order to random binary sequential string

   The memory aspect to the random binary sequential string is, in effect, the retaining of the number and order of the information found in the original pre-compression state. This can be done by assigning a relation to the subgroup that has a quality of reducing and then returning to the original state that can be done with the use of simple arithmetic. By assigning each sub-group in the random binary sequential string with a value of the multiplication of the amount found in that sub-group, a quantity is given that can be retained for use in reducing and expanding to the original size of that quantity and can be represented by a single character that represents the total number of characters found in that sub-group. This is the very nature of compression and duplicates the process found in the non-random binary sequential strings. As an example the random binary sequential string [110001001101111] can be grouped into sub-groups as follows: {11}, {000}, {1}, {00}, {11}, {0}, and {111} with each sub-group bracketed into common families of like digits. An expedient method to reduce this string would be to take similar types and reduce to a single character that represented a multiple of the exact number of characters found in that sub-group. In this case taking the bracketed {11} and assign a multiple of 2 to a single character, 1, and then reduced it to a single character in the bracket that is underlined to note the placement of the compression.

   The compressed random binary sequential string would appear like this: [1000100101111] with the total character length of 13, exhibiting the loss of two characters due to the compression of the two similar sub-groups. De-compression would be the removal of the underlining of each character and the replacement of the 1's characters to each of the sub-groups that would constitute a 100% retention of the original character number and order to the random binary sequential string. This makes for a new measure of kolmogorov Complexity in a random binary sequential string.

This same method of compression can be used with a radix 5 based system that provides for an even greater measure of reduction than is found in the binary sequential string. The radix 5 base number system has five separate characters that have no semantic meaning except not representing the other characters in the five character system. The following five numbers will represent the five characters found in the radix 5 base number system that will be used as an example in this paper: [0, 1, 2, 3 & 4]. As an example of a random radix 5 sequential string the following would appear like this: [001112233334440111223444] with a total character length of 24 characters. If all the applicable similar sequential 3 character's are compressed to a single representative character that represents the other two characters in the three character compressed unit of the string, then the following would result: [0012233334012234].

The underlined characters represent the initial position of the three character group of similar characters with a compressed state of 16 characters total. This is a reduction of one third the total original character length of 24 characters. A non-random radix 5 base sequential string will have the same character types: [0, 1, 2, 3 & 4] but with a regular pattern of groupings such as [00112233440011223344] that has a total character length of 20 and if all two sequentially similar characters are compressed using all 5 character types the following will occur: [0123401234] resulting in a compressed non-random radix 5 base sequential string of 10.

The paper has shown that a sub-maximal measure of Kolmogorov complexity exists that has implications to a new standard of the precise measure of randomness in both a radix 2 and a radix 5 based number systems.

References

[1] Kotz, S. and Johnson, N.I. Encyclopedia of Statistical Sciences (John Wiley & Sons, New York, 1982).

[2] abide.

[3] Solomonoff, R.J., Inf. & Cont. 7, 1-22 & 224-254 (1964), A.N. Kolmogorov, Pro. Inf. & Trans. 1, 1-7 (1965) and G.J. Chaitin, Jour. ACM 16, 145-159 (1969).

[4] Shannon, C.E., Bell Labs. Tech. Jour. 27, 379-423 and 623-656 (1948).

[5] Tice, B.S. "The use of a radix 5 base for transmission and storage of information", Poster for the Photonics West Conference, San Jose, California Wednesday January 23, 2008.

[6] Kotz, S. and Johnson, N.I. Encyclopedia of Statistical Sciences (John Wiley & Sons, New York, 1982).

Patterns Within Patternless Sequences[*]

Bradley S. Tice

While Kolmogorov complexity, also known as Algorithmic Information Theory, defines a measure of randomness as being pattern-less in a sequence of a binary string, such rubrics come into question when sub-groupings are used as a measure of such patterns in a similar sequence of a binary string. This paper examines such sub-group patterns and finds questions raised about existing measures for a random binary string.

PACS Numbers: 89.70tc, 89.20Ff, 89.70tc, 84.40Ua

Qualities of randomness and non-randomness have their origins with the work of von Mises in the area of probability and statistics [1]. While most experts feel all random probabilities are by nature actually pseudo-random in nature, a sub-field of statistical communication theory, also known as information theory, has developed a standard measure of randomness known as Kolmogorov randomness, also known as Martin-Lof randomness, that was developed in the 1960's [2,3 & 4]. This sub-field of information theory is known as Algorithmic Information Theory [5]. What makes this measure of randomness, and non-randomness, so distinct is the

---

*The paper was published by GRIN Publishing GmbH, Munich, Germany in 2012

notion of patterns, and pattern less, sequences
of 1's and 0's in a string of binary symbols [6].
In other words, perceptual patterns as seen in a
sequence of objects that can be defined as having
similar sub-groupings within the body of the
sequence that have a frequency, depending on the
length of the string, of either regularity, non-
randomness, or infrequency, randomness, within the
sequence itself [7].

In examining the classical notion of a random
and non-random set of 1's and 0's in two examples
of a sequence of binary strings, the pattern verses
pattern-less qualities can be examined. Example
#1 is as follows: [111000111000111] and Example
#2 is as follows: [110111001000011]. It is clear
than Example #1 is more patterned than Example
#2 in that Example #1 has a balanced sub-groups
of three characters, either all 1's or all 0's,
that have a perceptual regularity. Example #2 is a
classical model of a sequence of a random binary
string in that the sub-groups, if grouped into
like, or similar, characters, either all 1's or all
0's like in Example #1, the frequency of the types
of characters, either 1's or 0's, is different,
seven variations of groups as opposed to the five
variations in Example #1, as are the sub-groups:
[(11), (0), (111), (00), (1), (0000), & (11)] from
Example #2. While this would support the pattern
verses patternless model proposed by Kolmogorov
complexity, there is a striking result from these
two Examples, #1 and #2, in that the second, or
random, example, Example #2, has a pattern within
the sub-groups, that for all perceptual accounts,
has distinct qualities that can be used to measure
the nature of randomness on a sub-grouped level on
examination of a binary string.

The author has done early work on coding each of
the sub-groups and reducing them to a compressed

state, and then decompressing them with no loss to either the amount of frequency or number of characters to a sequence of a binary string that would be considered random by Kolmogorov complexity [8]. Now, while this simple program of compression and decompression by the author is for a future paper, the real interest of this paper is on the sub-groups as they stand without the notion of compression.

The very idea of the notion of a patterned or pattern-less quality as found in the measure of such aspects to the sub-groupings of 1's and 0's in a sequence of a binary string has the quality of being a bit vague, in that both Example #1 and Example #2 are patterned, in that they have a frequency and similar character sub-groupings that have a known measure and quality that can be quantified in both examples. This is more than a question of semantics as the very nature of the measure of Kolmogorov complexity is the very fact that it has a perceptual 'pattern' to measure the randomness of a sequence of a binary string. In reviewing the literature on the notions of patterns in Kolmogorov complexity/Algorithmic Information Theory the real question arises, which patterns qualify for status as random, especially as a measure in a sequence of a binary string?

References

[1] Knuth, D.E., The Art of Computer Programming: Volume 2 Semi numerical Algorithms (Addison-Wesley Publishers, Reading), 1997, p. 149.

[2] Knuth, D.E., The Art of Computer Programming: Volume 2 Semi numerical Programming (Addison-Wesley Publishers, Reading), 1997, p. 169-170.

[3] Shannon, C.E., Bell Labs. Tech. Jour. 27, (1948), 379-423 & 623-656.

[4] Li, M. and Vitanyi, P., An Introduction to Kolmogorov Complexity and Its Applications (Springer, New York), 1997, p. 186.

[5] Ge, M., The New Encyclopedia Britannica (Encyclopedia Britannica, Chicago), 2005, p. 637.

[6] Martin-Lof, P., Infor. And Contr., 9.6 (1966), 602–619.

[7] Uspensky, V.A., 'An introduction to the theory of kolmogorov complexity' edited by Watanabe, O. Kolmogorov Complexity and Computational Complexity (Springer-Verlag, Berlin), 1992, p. 87.

[8] Tice, B.S., Formal Constraints to Formal Languages (Author House, Bloomington), in press.

A Radix 4 Base System for Use in Theoretical Genetics

Poster #7888-27 accepted and scheduled for presentation for the "Frontiers in Biological Detection: From Nanoscience to Systems". SPIE BIOS Conference, January 23, 2011, San Francisco, California U.S.A.

A Radix 4 Based System for Use in Theoretical Genetics

By Bradley S. Tice

ABSTRACT

The paper will introduce the quaternary, or radix 4, based system for use as a fundamental standard beyond the traditional binary, or radix 2, based system in use today. A greater level of compression is noted in the radix 4 based system when compared to the radix 2 base as applied to a model of information theory. The application of this compression algorithm to both DNA and RNA sequences for compression will be reviewed in this paper.

Keywords: Radix 4, Quaternary, Theoretical Genetics, DNA Compression, RNA Compression

1. Introduction

A quaternary, or radix 4 based system, is defined as four separate characters, or symbols, that have no semantic meaning apart from not representing the other characters. This is the same notion Shannon gave to the binary based system upon it's publication in 1948 [1]. This paper will present research that shows the radix 4 based system to have a compression value greater than the traditional radix 2 based system in use today [2]. The compression algorithm will be used to compress DNA and RNA sequences. The work has applications in theoretical genetics and synthetic biology.

2. Randomness

The earliest definition for randomness in a string of 1's and 0's was defined by von Mises, but it was Martin-Lof's paper of 1966 that gave a measure to randomness by the *patternlessness* of a sequence of 1's and 0's in a string that could be used to define a random binary sequence in a string [3 and 4]. A non-random string will be able to compress, were as a random string of characters will not be able to compress. This is the classical measure for Kolmogorov complexity, also known as Algorithmic Information Theory, of the randomness of a sequence found in a binary string.

3. Compression Program

The compression program to be used has been termed the *Modified Symbolic Space Multiplier Program* as it simply notes the first character in a line of characters in a binary sequence of a string and subgroups them into common or like groups of similar characters, all 1's grouped with 1's and all 0's grouped with 0's, in that string and is assigned a single character notation that represents the number found in that sub-group, so that it can be reduced, compressed, and decompressed, expanded, back to it's original length and form [5]. An underlined 1 or 0 is usually used to note the notation symbol for the placement and character type in previous applications of this program. The underlined initial character to be compressed will be used for this paper.

4. Application of Theory

The application of a quaternary, or radix 4 based system, to existing genetic marking and counting systems has many advantages. The first is the greater amount of compression from this base, as opposed to the standard binary based system in use today, and secondly, as a more utilizable system because of the four character, or symbol, based system that provides for more variety to develop information applications.

5. DNA

DNA, or Deoxyribonucleic acid, is a linear polymer made up of specific repeating segments of phosphodiester bonds and is a carrier of genetic information [6]. There are four bases in DNA; adenine, thymine, guanine and cytosine [7].

The use of a compression algorithm for sequences of DNA.

Definitions
A = Adenine
T = Thymine
G = Guanine
C = Cytosine

Example #A

ATATGCGCTATACGCGTATATATA

The compression algorithm will use a specific focus on TA and GC DNA sequences in Example #A.

Key Code

TA = 4 characters
GC = 2 characters

Compress Example #A

ATAT<u>GC</u>ATATCGCG<u>TA</u>

The compressed DNA sequence is 16 characters from the original non-compression total of 24.

The use of a four character system, a radix 4 base number system, that is composed of each character not representing the other characters is ideal in DNA sequences composed of adenine, thymine, guanine and cytosine.

Example #D

TAGCTAGCTAGCTAGCTAGCTAGCTAGCTAGCTAGCTAGC

Key Code

TAGC = 10

Compression of Example #D

TAGC

The compressed version of Example #D is 4 characters from the original non-compressed total of 40 characters.

6. RNA

RNA, or Ribonucleic acid, translates the genetic information found in DNA into proteins [8]. There are four bases that attach to each ribos [9].

The use of a compression algorithm for sequences of RNA.

Definitions
A = Adenine
C = Cytosine
G = Guanine
U = Uracil

Example #B

AUAUCGCGAUAUCGCGUAUAUAUAGCGC

The compression algorithm will focus on specific RNA sequences.

Key Code

UA = 4 characters

GC = 2 characters

Compress Example #B

AUAUCGCGAUAUCGCG<u>UAGC</u>

   The compressed RNA sequence is 20 characters in length form the original non-compression total character length of 28.

   The use of a four character system, a radix 4 base number system, that is composed of each character not representing the other characters is ideal in RNA sequences composed of adenine, cytosine, guanine and uracil.

Example #C

UAGCUAGCUAGCUAGCUAGCUAGC

The use of a universal compression algorithm is as follows:

Key Code

UAGC = 6

Compression of Example #C

<u>UAGC</u>

   The compressed version of Example #C is 4 characters from the original non-compressed 24 character total length.

Summary

The compression algorithm used for both DNA and RNA sequences has the power of both a universal compression algorithm, all character length types, and a specific, or target, level of compression.

References

[1] "Shannon, C.E., "A Mathematical Theory of Information", *Bell Labs. Tech. Jour.* 27, 379-423 and 623-656 (1948).

[2] Tice, B.S., "The analysis of binary, ternary and quaternary based systems for communications theory", Poster for the SPIE Symposium on Optical Engineering and Application Conference, San Diego, California, August 10-14, 2008.

[3] Kotz, S. and Johnson, N.I., Encyclopedia of Statistical Sciences, John Wiley & Sons, New York (1982).

[4] Martin-Lof, P., "The definition of random sequences", *Information and Control*, 9, pp. 602-619 (1966).

[5] Tice, abide.

[6] Lutter, L.C. "Deoxyribonucleic acid". In *McGraw-Hill Encyclopedia of science & technology*. McGraw-Hill Publishers, New York, pp. 373-379 (2007).

[7] Lutter, abide., p. 374.

[8] Beyer, A.L. and Gray, M.W. "Ribosomes". In *McGraw-Hill Encyclopedia of science & technology*. McGraw-Hill Publishers, New York, pp. 542-546 (2007).

[9] Beyer, abide., p. 542.

A Compression Program for Chemical, Biological and Nanotechnologies

Poster #7910-58 accepted and scheduled for presentation for the "Reports, Markers, Dyes, Nanoparticles, and Molecular Probes for Biomedical Applications". SPIE BIOS Conference, January 24, 2011, San Francisco, California U.S.A.

A Compression Program for Chemical, Biological, and Nanotechnologies

By Bradley S. Tice

ABSTRACT

The paper will introduce a compression algorithm that will use based number systems beyond the fundamental standard of the traditional binary, or radix 2, based system in use today. A greater level of compression is noted in these radix based number systems when compared to the radix 2 base as applied to a sequential strings of various information. The application of this compression algorithm to both random and non-random sequences for compression will be reviewed in this paper. The natural sciences and engineering applications will be areas covered in this paper.

Keywords: Compression Algorithm, Chemistry, Biology, and Nanotechnology

1. Introduction

A binary, or radix 2 based, system is defined as two separate characters, or symbols, that have no semantic meaning apart from not representing the other character. This is the same notion Shannon gave to the binary based system upon it's publication in 1948 [1]. This paper will present research that shows how various radix based number systems have a compression value greater than the traditional radix 2 based system as in use today

[2]. The compression algorithm will be used to compress various random and non-random sequences. The work has applications in theoretical and applied natural sciences and engineering.

2. Randomness

The earliest definition for randomness in a string of 1's and 0's was defined by von Mises, but it was Martin-Lof's paper of 1966 that gave a measure to randomness by the *patternlessness* of a sequence of 1's and 0's in a string that could be used to define a random binary sequence in a string [3 and 4]. A non-random string will be able to compress, were as a random string of characters will not be able to compress. This is the classical measure for Kolmogorov complexity, also known as Algorithmic Information Theory, of the randomness of a sequence found in a binary string.

3. Compression Program

The compression program to be used has been termed the *Modified Symbolic Space Multiplier Program* as it simply notes the first character in a line of characters in a binary sequence of a string and subgroups them into common or like groups of similar characters, all 1's grouped with 1's and all 0's grouped with 0's, in that string and is assigned a single character notation that represents the number found in that sub-group, so that it can be reduced, compressed, and decompressed, expanded, back to it's original length and form [5]. An underlined 1 or 0 is usually used to note the notation symbol for the placement and character type in previous applications of this program. The underlined initial character to be compressed will be used for this paper.

4. Application of Theory

The compression algorithm will be used for the following radix based number systems: Radix 6, Radix 8, Radix 10, radix 12 and radix 16. These are traditional radix base numbers from the field of computer science and have strong applications to other fields of science and engineering due to the parsimonious nature of these low digit radix base number systems [6]. The compression algorithm in this paper can be both a 'universal' compression engine in that all members of a sequence, either random or non-random, can be compressed or a 'specific' compression engine that compresses only specific types of sub-groups within a random or non-random string of a sequence.

The compression algorithm will be defined by the following properties:

1) Starting at the far left of the string, the beginning, and moving to the right, towards the end of the string.
2) Each sub-group of common characters, including singular characters, will be grouped into common sub-groups and marked accordingly.
3) The notation for marking each sub-group will be underling the initial character of that common sub-group. The remaining common characters in that marked sub-group will be removed. This results in a compressed sequential string.
4) De-compression of the compressed string is the reverse process with complete position and character count to the original pre-compressed sequential string.
5) This will be the same processes for both random and non-random sequential strings.

5. Chemistry

Chemistry is the science of the structure, the properties and the composition of matter and it's changes [7].

5.1 Polymer

A polymer is macromolecule, large molecule, made up of repeating structural segments usually connected by covalent chemical bonds [8].

5.2 Copolymer

A copolymer, also known as a heteropolymer, is a polymer derived from two or more monomers [9].

Types of Copolymers;

1) Alternating Copolymers: Regular alternating A and B units.
2) Periodic Copolymers: A and B units arranged in a repeating sequence.
3) Statistical Copolymers: Random sequences.
4) Block Copolymers: Made up of two or more homopolymer subunits joined by covalent bonds.
5) Stereoblock Copolymer: A structure formed from a monomer.

An example of the use of a compression algorithm on copolymers is as follows:

1) Alternating Copolymers: Alternating copolymers using a radix 2 base number system.
Unit A = 0
Unit B = 1

Example #1:

01010101010101
Compression of Example #1

Key Code

0 = 7 characters
1 = 7 characters

Example #1 Compressed

<u>01</u>
   The compressed state of Example #1 is a 2 character length from the original non-compression state total of 17 characters in length.

Periodic Copolymers: Periodic copolymers using a radix 16 base number system.

Unit A = abcdefghijklmnop
Unit B = 123456789@#$%^&*

Example #2

abcdefghijklmnop123456789@#$%^&*123456789@#$%^&*a
bcdefghijklmnop123456789@#$%^&*

Compression of Example #2

Key Code

abcdefghijklmnop = 16 characters
123456789@#$%^&* = 16 characters

Example #2 Compressed

<u>allal</u>
   The compressed state of Example #2 is 5 characters from the original non-compression state total of a 80 character length.

   Statistical Polymers: Random copolymer using a radix 8 base number system.

Unit A = 12345678
Unit B = abcdefgh

Example #3

12345678abcdefghabcdefgh12345678
12345678abcdefgh12345678123456781234567812345678

Key Code

12345678 = 8 characters
abcdefgh = 8 characters

Compression of Example #3

lallalll

The compressed state of Example #3 is 8 from the original non-compression state total of a 64 character length.

 Block Copolymers: Block copolymer using a radix 12 base number system.
Unit A = abcdefghijkl
Unit B = 123456789@#$

Example #4

123456789@#$123456789@#$123456789@#$abcdefghijkla
bcdefghijklabcdefghijkl

Key Code

abcdefghijk = 12 characters
123456789@#$ = 12 characters

Compression of Example #4

lllaaa

 The compressed state of Example #4 is 6 characters from the original non-compression state of 58 character length.

 Stereoblock Copolymer: Stereoblock copolymer using a radix 10 base number system.

```
Unit A = abcdefghij
Unit B = 123456789@
```

Note: The symbol [I] represents a special structure defining each block.

Example #5

```
abcdefghijabcdefghijabcdefghijabcdefghijabcdefgh
ijabcedfghij
              I                          I
123456789@123456789@        123456789@123456789@
```

Key Code

```
abcedfghij = 10 characters
123456789@ = 10 characters
```

Compression of Example #5

```
aaaaaa
1 1 1 1
```

The compressed state of Example #5 is 10 characters from the original non-compression total of 100 characters in length.

6. Biology

Biology is the study of nature and as such is a part of the systematic atomistic axiomization of processes found within living things. These natural grammars, or laws, has mathematical corollates that parallel process found in the physical and engineering disciplines. The use of a compression algorithm of a sequential string is a natural development of such a process as can be seen in the compression of both DNA and RNA genetic codes.

6.1 DNA

DNA or Deoxyribonucleic acid, is a linear polymer made up of specific repeating segments of phosphodiester bonds and is a carrier of genetic information [10]. There are four bases in DNA; adenine, thymine, guanine and cytosine [11].

The use of a compression algorithm for sequences of DNA.

Definitions:
A = Adenine
T = Thymine
G = Guanine
C = Cytosine

Example #A

ATATGCGCATATCGCGTATATATATATA

The compression algorithm will use a specific focus on TA and GC DNA sequences in Example #A.

Key Code

TA = 6 characters
GC = 2 characters

Compress Example #A

ATAT<u>GC</u>ATATCGCG<u>TA</u>

The compressed DNA sequence is 16 characters from the original non compression total of a 28 character length.

6.2 RNA

RNA, or Ribonucleic acid, translates the genetic information found in DNA into proteins [12]. There are four bases that attached to each ribos [13].

Definitions:
A = Adenine
C = Cytosine
G = Guanine
U = Uracil

Example #B

AUAUCGCGAUAUCGCGUAUAUAUAUAUAGCGC

The compression algorithm will focus on specific RNA sequences.

Key Code

UA = 6 characters
GC = 2 characters

Compress Example #B

AUAUCGCGAUAUCGCGUAGC

The compressed RNA sequence is 20 characters in length from the original non-compression total character length of 32.

7. Nanotechnology

The development and discovery of nanometer scale structures, ranging from 1 to 100 nanometers, to transform matter, energy and information on a molecular level of technology [14].

7.1 Synthetic Biology

Within the field of synthetic biology is the development of synthetic genomics that uses aspects of genetic modification on pre-existing life forms to produce a product or desired behavior in the life form created [15].

The following is a DNA sequence of real and 'made up' synthetic sequences.

Definitions:
A = Adenine
T = Thymine
G = Guanine
C = Cytosine
W = *Watson
K = *Crick
*Note: Made up synthetic DNA.

Example #C

TATAGCGCWKWKATATCGCGKWKWKWKWKWKW

Key Code

AT = 2 characters
CG = 2 characters
KW = 6 characters

Compressed Example #C

TATAGCGCWKWKATCGKW

The compressed synthetic DNA sequence is 18 characters from the original non-compression character total of 32.

Summary

The paper has addressed the use of a compression algorithm for use in various radix based number systems in the fields of chemistry, biology and nanotechnology. The compression algorithm in both the universal and specific format have successfully reduced long and short sequences of strings to very compressed states and function well in both random and non-random sequential strings.

References

[1] Shannon, C.E., *Bell Labs. Tech. Jour.* 27, 379–423 and 623–656 (1948).

[2] Tice, B.S., "The analysis of binary, ternary and quaternary based systems for communications theory", Poster for the SPIE Symposium on Optical Engineering and Application Conference, San Diego, California, August 10-14, 2008.

[3] Kotz, S. and Johnson, N.I., *Encyclopedia of Statistical Sciences*, John Wiley & Sons, New York (1982).

[4] Martin-Lof, P., "The definition of random sequences", *Information and Control*, 9, pp. 602–619 (1966).

[5] Tice, abide.

[6] Richards, R.K., *Algorithmic Operations in Digital Computers*, D. Van Nostrand Company, Princeton, N.J., (1955).

[7] Moore, J.A., *McGraw-Hill Encyclopedia of Chemistry*, McGraw-Hill Publishers, New York (1993).

[8] Wikipedia, "Polymer". Wikipedia, September 4, 2010, p. 1. Website: http://en.wikipedia.org/wiki/Polymers.

[9] Wikipedia, "Copolymer". Wikipedia, September 4, 2010, pp. 1–5. Website: http://ep.wikipedia.org/wiki/Copolymers.

[10] Lutter, L.C., "Deoxyribonucleic acid". In *McGraw-Hill Encyclopedia of science & technology*. McGraw-Hill Publishers, New York, pp. 373–379 (2007).

[11] Lutter, abide., p. 374.

[12] Beyer, A.L. and Gray, M.W., "Ribosomes". In *McGraw-Hill Encyclopedia of science & technology*. McGraw-Hill Publishers, New York. pp. 542-546 (2007).

[13] Beyer, abide., p. 542.

[14] Drexler, K.E., "Nanotechnology". In *McGraw-Hill Encyclopedia of science & technology*. McGraw-Hill Publishers, New York, pp. 604-607 (2007).

[15] Wikipedia. "Synthetic genomics". Wikipedia, September 4, 2010, p. 1. Website: http://ep.wikipedia.org/wiki/Synthetic-genomics.

Babbage, Enigmas and Captain Crunch— An Essay

Bradley S. Tice

This essay is connected by stories of calculating machines, cryptology and lastly the joining of the two areas in the form of 20th Century computer hacking. A unifying experience for my personal life and family life is joined by these connections.

I have chosen the engineering drawings from Charles Babbage's (1791–1871) mechanical calculator, the Analytical Engine, as a stylistic metaphor for my algorithmic compression program, a compression engine, that gives the esthetics of the mechanical properties of my algorithm with the tangible aspects of 19th Century mechanical calculating machines.

The Enigma machine was a top secret German coding machine from World War Two and were used to encipher and de-cyther secret codes. A German army and a German naval version were used during the war (Kahn, 1991: 43). I was able to get a close hand look at an Enigma machine at the Computer Museum in Mountain View, California U.S.A. some years ago as recounted in the Notes section of this paper (Note #1).

In 1971 my late father moved our family from Cupertino, California to the graduate family housing of Stanford University, known as Escondido Village.

Because he needed a private mail box on campus he procured a P.O. box on campus at the time and was soon receiving the previous occupants mail. Of these a few letters stood out as they were addressed to 'Captain Crunch' of breakfast cereal fame. When my father turned this mail into the resident Postmaster, he asked him about the Captain Crunch letter's. The Postmaster informed him that it was the name of the notorious hacker of telephone lines that used the toy whistle in the cereal box to obtain 'free' telephone calls around the world. After a period of time the letters stopped and Captain Crunch faded into computer hacking history (Note #2).

This rather disjointed series of events has found a pattern reflected in this monograph on A Level of Martin-Lof Randomness that combines my father's graduate school experience with my museum, the Enigma machine, and archival experience, technical drawings by Charles Babbage, as well as seeing Babbage's work at the Science Museum in London, England, U.K. over a period of years during my travels.

The final result is an esthetic visual complement to the overtly mechanistic aspects inherent to an algorithm and especially to the compression engine as detailed in this monograph.

Notes

Note #1: I would like to thank Mr. Leonard J. Shustek, Chairman of the Board of the Computer History Museum, Mountain View, California U.S.A. for giving me a personal tour of the museum, it had just opened, in 2003 and allowed me to operate their Enigma machine that had been connected to a personal computer by an engineer from Sun Microsystems.

Note #2: Steve Wozniak, the co-founder of Apple Computer, has retold his experience with Captain Crunch in his biography, iWoz (2006).

References

Kahn, David (1991) Seizing the Enigma: The Race to Break the German u-Boat Codes 1939-1943 (New York: Barnes and Noble).

Wozniak, Steve and Smith, Gina (2006) iWoz (New York: W.W. Norton & Company).

Innovation and Mathematics

Bradley S. Tice

January 28, 2011

ABSTRACT

The paper will address the question of innovation and it's role in developing mathematical intuition and invention in the development of robust aspects to my compression algorithm, a universal application, but also a truncated, or partial, compression of a random sequential string. As the discovery of the compression algorithm was by 'chance' and rather a capricious happenstance of events, the secondary aspects of this compressible algorithm is the compressible natures of both random properties of all of the sequential string, or just parts of the sequential string. This was the bases for this papers' premise.

Introduction

While the discovery of compression ratios lower than the original non-compressed states in random sequential binary strings was by happenstance, the examined and tested aspects to this statistical novelty was based on scientific techniques and the mathematical philosophy of 'plausible inference' as was expressed by the mathematician George Polya (1887-1985) [Note #1].

Universal and Specific Aspects to a Sequential String

What makes the algorithm for a sub-maximal level of compression in a sequential string is that the program can engage all aspects of a sequential string, a universal feature of the program, but also very specific types of features, character types, found in the sequential string. This makes the compression algorithm multifaceted as it can be both a total system compression or a specific, truncated, system of compression. While the examination and testing of this compression algorithm was done at the abstract level, the practical aspects to this system to both engineering and computer science became obvious (Tice, 2009).

Mathematical Induction

Polya considers a scientists experience to be a form of induction, of which mathematics offers the best examples of such powers (Polya, 1954: 4). A 'suggestive contact' is one that begins with an observation, but notes Polya, one should not be swayed by conjecture alone, unsupported claims, as it must have 'supporting contact'; those aspects that verify a conjecture beyond the initial observation (Polya, 1954: 7).

Polya makes the salient comment that to be a good mathematician, you must be a 'good guesser', and to be a good guesser one should have certain aptitudes that make the process of guessing an 'educational' one that serves the mathematician better at making future guesses (Polya, 1954: 112).

When I first observed the compression of a random sequential binary string, I was only interested in a total compression of the string, specific aspects of this algorithm were of secondary importance to the very fact that the random sequential string generated at a compressed state was lower than known values for a random binary sequential string (Tice, 2009: 48). In time, I began to examine other aspects to the algorithms program and to test the practical limits to such a system. It was clear, experience or 'induction', was going to play a role in trying to test parts of a string, either the 1 or the 0 in a binary system, and sure enough, focusing on either the 1 or the 0 in a binary sequential string resulted in all those specific characters, either the 1 or the 0, to compress as per dictated by the compression algorithm.

With the examination and testing of larger radix base number systems beyond the binary level, more characters could be used to test the specific nature of compression of both random and non-random sequential strings. Again, experimental proof was needed to support my conjecture that the compression algorithm could be utilized for larger radix base number systems. By testing these larger radix base number systems at both the random and non-random level, a fundamental level of compression abilities and limits could be empirically judged and validated.

Polya on Randomness

In one of polya's books, he examines the role of statistical probability and randomness as 'observable phenomena' of our natural world (Polya, 1968: 55). Polya uses the example of rain drops falling on two stones in a open space being hit by rain drops (Polya, 1968: 56). Recording each rain drop to fall on each stone, one a L or Left stone, the other the Right stone or R, and the following rain drop distribution is recorded as a sequential string of L and R raindrop hits: LRRLLLRLRLRRLRR

(Polya, 1968: 56). From this Polya states that while it is highly probably that both stones will get completely wet, which stone, L or R, will be hit next after the 15th recorded rain drop is unknown (Polya, 1968: 56). As Polya contends from this experiment: we can foresee the long term, both stones getting totally wet, but can not foresee the details of such phenomena, the number of rain drops to hit either the Left stone or the Right stone by the end of the rain storm (Polya, 1968: 56). This is the very nature of randomness as it is observed in the natural world.

Summary

While the discovery of compressible random sequential strings was a 'eureka' moment, the extent and types of sequential string compression at the random and non-random levels only came about through examination and testing of the compression algorithm and being able to be judged and validated on an empirical level. The level of experience and 'accumulated induction' made this process possible for the author to undertake such systematic testing and validating of the compression algorithm.

References

Polya, George (1954) Induction and Analogy in Mathematics. Princeton: Princeton University Press.

Polya, George (1968) Patterns of Plausible Inference. Princeton: Princeton University Press.

Tice, Bradley S. (2009) Aspects of Kolmogorov Complexity: The Physics of Information. Denmark: River Publishers.

Young, Robyn V. (1998) Notable Mathematicians: From Ancient Times to the Present. New York: Gale, pp. 401-403.

Notes

I can remember Dr. George Polya stopping by on his weekly rounds to and from the Stanford University campus to admire my mother's small flower garden she had developed while we were living in the graduate family housing, Escondido Village, while my father pursued his graduate degree in computer science and mathematics, 1971–1975. I know this from my father's amazement one day when he saw Dr. Polya talking to my mother in front of the apartment about her plants. My mother did not know who the 'nice old gentleman' was and was interested to learn that he was one of the most respected mathematicians in the world*.

*Citation: R.V. Young (Editor) (1998) Notable Mathematicians: From Ancient Times to the Present. New York: Gale, pp. 401–403.

Raymond J. Solomonoff
(July 25, 1926–December 7, 2009)—An Obituary

Peter Gács and Paul M.B. Vitányl

Ray Solomonoff, the first inventor of some of the fundamental ideas of Algorithmic Information Theory, died in December, 2009. His original ideas helped start the thriving research areas of algorithmic information theory and algorithmic inductive inference. His scientific legacy is enduring and important. He was also a highly original, colorful personality, warmly remembered by everybody whose life he touched. We outline his contributions, placing it into its historical context, and the context of other research in algorithmic information theory.

1. Introduction

Raymond J. Solomonoff died on December 7, 2009, in Cambridge, Massachusetts. He was the first inventor of some of the fundamental ideas of Algorithmic Information Theory, which deals with the shortest effective description length of objects and is commonly designated by the term "Kolmogorov complexity."

In the 1950s Solomonoff was one of the first researchers to introduce probabilistic grammars and the associated languages. He championed probabilistic methods in Artificial Intelligence

(AI) when these were unfashionable there, and treated questions of machine learning early on. But his greatest contribution is the creation of Algorithmic Information Theory.

In November 1960, Solomonoff published the report [14] presenting the basic ideas of Algorithmic Information Theory as a means to overcome serious problems associated with the application of Bayes's rule in statistics. His findings (in particular, the invariance theorem) were mentioned prominently in April 1961 in Minsky's symposium report [8]. (Andrei N. Kolmogorov, the great Russian mathematician, started lecturing on description complexity in Moscow seminars about the same time.)

Solomonoff's objective was to formulate a completely general theory of inductive reasoning that would overcome shortcomings in Carnap's [1]. Following some more technical reports, in a long journal paper in two parts he introduced "Kolmogorov" complexity as an auxiliary concept to obtain a universal a priori probability and proved the invariance theorem that, in various versions, is one of the characteristic elements of Algorithmic Information Theory [16,17]. The mathematical setting of these ideas is described in some detail below.

Solomonoff's work has led to a novel approach in statistics leading to applicable inference procedures such as the minimal description length principle. Jorma J. Rissanen, credited with the latter, relates that his invention is based on Solomonoff's work with the idea of applying it to classical statistical inference [10,11].

Since Solomonoff is the first inventor of Algorithmic Information Theory, one can raise the question whether we ought to talk about "Solomonoff complexity." However, the name "Kolmogorov complexity" for shortest effective description length has become well entrenched and is commonly understood. Solomonoff's publications apparently received little attention until Kolmogorov started to refer to them from 1968 onward. Says Kolmogorov, "I came to similar conclusions [as Solomonoff], before becoming aware of Solomonoff's work, in 1963–1964" and "The basic discovery, which I have accomplished independently from and simultaneously with R. Solomonoff, lies in the fact that the theory of algorithms enables us to eliminate this arbitrariness by the determination of a 'complexity' which is almost invariant (the replacement of one method by another leads only to the addition of a bounded term)".

Solomonoff's early papers contain in veiled form suggestions about randomness of finite strings, incomputability of Kolmogorov complexity computability of approximations to the Kolmogorov complexity, and resource-bounded Kolmogorov complexity.

Kolmogorov's later introduction of complexity was motivated by information theory and problems of randomness. Solomonoff introduced algorithmic complexity independently and earlier and for a different reason: inductive reasoning. Universal a priori probability, in the sense of a single prior probability that can be substituted for each actual prior probability in Bayes's rule was invented by Solomonoff with Kolmogorov complexity as a side product, several years before anybody else did.

A third inventor is Gregory J. Chaitin, who formulated a proper definition of Kolmogorov complexity at the end of his paper [2].

For a more formal and more extensive study of most topics treated in this paper, we recommend [7].

2. The Inventor

Ray Solomonoff published a scientific autobiography up to 1997 as [23]. He was born on July 25, 1926, in Cleveland, Ohio, in the United States. He studied physics during 1946–1950 at the University of Chicago (he recalls the lectures of E. Fermi). He obtained a Ph.B. (bachelor of philosophy) and a M.Sc. in physics. He was already interested in problems of inductive inference and exchanged viewpoints with the resident philosopher of science at the University of Chicago, Rudolf Carnap, who taught an influential course in probability theory.

From 1951–1958 he held half-time jobs in the electronics industry doing math and physics and designing analog computers.

In 1956, Solomonoff was one of the 10 or so attendees of the Dartmouth Summer Research Conference on Artificial Intelligence, at Dartmouth College in Hanover, New Hampshire, organized by M. Minsky, J. McCarthy and C.E. Shannon, and in fact stayed on to spend the whole summer there. (This meeting gave AI its name.) There Solomonoff wrote a memo on inductive inference.

McCarthy had the idea that given every mathematical problem, it could be brought into the form of "given a machine and a desired output, find an input from which the machine computes that output." Solomonoff suggested that there was a class of problems that was not of that form: "given an initial segment of a sequence, predict its continuation." McCarthy then thought that if one saw a machine producing the initial segment, and then continuing past that point, would one not think that the continuation was a reasonable extrapolation? With that the idea got stuck, and the participants left it at that.

Also in 1956, Ray circulated a manuscript of "An Inductive Inference Machine" at the Dartmouth Summer Research Conference on Artificial Intelligence, and in 1957 he presented a paper with the same name at the IRE Convention, Section on Information Theory, a forerunner of the IEEE Symposium on Information Theory. This partially used Chomsky's paper [3] read at a Symposium on Information Theory held at MIT in September 1956. "An Inductive Inference Machine" already stressed training sequences and using previous solutions in solving more complex problems. In about 1958 he left his half-time position in industry and joined Zator Company full time, a small research outfit located in some rooms at 140 1/2 Mount Auburn Street, Cambridge, Massachusetts, which had been founded by Calvin Mooers around 1954 for the purpose of developing information retrieval technology. Floating mainly on military funding, Zator Co. was a research front organization, employing Mooers, Solomonoff, Mooers's wife, and a secretary, as well as at various times visitors such as Marvin Minsky. It changed its name to the more martial sounding Rockford Research (Rockford, Illinois, was a place where Mooers had lived) around 1962. In 1968, the US Government reacted to public pressure (related to the Vietnam War) by abolishing defense funding of civil research, and Rockford foundered. Being out of a job, Solomonoff left and founded his own (one-man) company, Oxbridge Research, in Cambridge in 1970, and has been there ever since, apart from spending nine months as research associate at MIT's Artificial Intelligence Laboratory, the academic year 1990–1991 at the University of Saarland, Saarbruecken, Germany, and a more recent sabbatical at IDSIA, Lugano, Switzerland.

It is unusual to find a productive major scientist that is not regularly employed at all. But from all the elder people (not only scientists) we know, Ray

Solomonoff was the happiest, the most inquisitive, and the most satisfied. He continued publishing papers right up to his death at 83.

In 1960 Solomonoff published [14], in which he gave an outline of a notion of universal a priori probability and how to use it in inductive reasoning (rather, prediction) according to Bayes's rule. This was sent out to all contractors of the Air Force who were even vaguely interested in this subject. In [16,17], Solomonoff developed these ideas further and defined the notion of enumeration, a precursor of monotone machines, and a notion of universal a priori probability based on his variant of the universal monotone machine. In this way, it came about that the original incentive to develop a theory of algorithmic information content of individual objects was Solomonoff's invention of a universal a priori probability that can be used as a priori probability in applying Bayes's rule.

Solomonoff's first approach was based on Turing machines with markers that delimit the input. This led to awkward convergence problems with which he tried to deal in an ad-hoc manner. The young Leonid A. Levin (who in [27] developed his own mathematical framework, which became the source of a beautiful theory of randomness), was told by Kolmogorov about Solmonoff's work. He added a reference to it, but had in fact a hard time digesting the informalities; later though, he came to appreciate the wealth of ideas in [16]. Solomonoff welcomed Levin's new formalism with one exception: it bothered him that the universal a priori probability for prediction is a semimeasure but not a measure (see below). He continued to advocate a normalization operation keeping up a long technical argument with Levin and Solovay.

In 2003 he was the first recipient of the Kolmogorov Award by The Computer Learning Research Center at the Royal Holloway, University of London, where he gave the inaugural Kolmogorov Lecture. Solomonoff was a visiting Professor at the CLRC. A list of his publications (published and unpublished) is at http://world.std.com/~rjs/pubs.html.

3. The Formula

Solomonoff's main contribution is best explained if we start with his inference formula not as he first conceived it, but in the cleaner form as it is known today, based on Levin's definition of apriori probability [27]. Let T be a computing device, say a Turing machine. We assume that it has some, infinitely expandable, internal memory (say, some tapes of the Turing machine). At each step, it may or may not ask for some additional input symbol from the alphabet $\{0,1\}$, and may or may not output some symbol from some finite alphabet Σ. For a finite or infinite binary string p, let $T(p)$ be the (finite or infinite) output sequence emitted while not reading beyond the end of p. Consider the experiment in which the input is an infinite sequence of tosses of an independent unbiased coin. For a finite sequence $x = x_1 \ldots x_n$ written in the alphabet Σ, let $M_T(x)$ be the probability that the sequence outputted in this experiment begins with x. More formally, let $T^{-1}(x)$ be the set of all those binary sequences p that the output string $T(p)$ contains x as a prefix, while if p' is a proper prefix of p then $T(p')$ does not output x yet. Then

$$M_T(x) = \sum_{p \in T^{-1}(x)} 2^{-|p|} \tag{1}$$

where $|p|$ is the length of the binary string p. The quantity $M_T(x)$ can be considered the *algorithmic probability* of the finite sequence x. It depends, of course, on the choice of machine T, but if T is a universal machine of the type called *optimal* then this dependence is only minor. Indeed, for an optimal machine U, for all machines T there is a finite binary r_T with the property $T(p) = U(r_T p)$ for all p. This implies $M_U(x) \geq 2^{-|r_T|}M_T(x)$ for all x. Let us fix therefore such an optimal machine U and write $M(x) = M_U(x)$. This is (the best-known version of) Solomonoff's *apriori probability*.

Now, Solomonoff's prediction formula can be stated very simply. Given a sequence x of experimental results, the formula

$$\frac{M(xy)}{M(x)} \qquad (2)$$

assigns a probability to the event that x will be continued by a sequence (or even just a symbol) y. In what follows we will have opportunity to appreciate the theoretical attractiveness of the formula: its prediction power, and its combination of a number of deep principles. But let us level with the reader: it is incomputable, so it can serve only as an ideal embodiment of some principles guiding practical prediction. (Even the apriori probability $M(x)$ by itself is incomputable, but it is at least approximable by a monotonic sequence from below.)

4. First, Informal Ideas

Scientific ideas of great originality, when they occur the first time, rarely have the clean, simple form that they acquire later. Nowadays one introduces description complexity ("Kolmogorov" complexity) by a simple definition referring to Turing machines. Then one proceeds to a short proof of the existence of an optimal machine, further to some simple upper and lower bounds relating it to probability and information. This a highly effective, formally impeccable way to introduce an obviously interesting concept.

Inductive inference is a harder, more controversial issue than information and randomness, but this is the problem that Solomonoff started with! In the first papers, it is easy to miss the formal definition of complexity since he uses it only as an auxiliary quantity; but he did prove the machine independence of the length of minimal codes.

The first written report seems to be [14]. It cites only the book [1] of Carnap, whose courses Solomonoff attended. And Carnap may indeed have provided the inspiration for a probability based on pure logical considerations. The technical report from allowed the gradual, informal development of ideas.

The work starts with confining the considerations to one particular formal representation of the general inference problem: predicting the continuations of a finite sequence of characters. Without making any explicit references, it sets out to combine two well-studied principles of inductive inference: Bayesian statistics and the principle that came to be known (with whatever historic justification) as "Occam's Razor". A radical version of this principle says that we should look for a shortest explanation of the experimental results and use this explanation for prediction of future experiments. In the context of prediction, it will be therefore often justified to call descriptions *explanations.*

Here is the second paragraph of the introduction:

> Consider a very long sequence of symbols—
> e.g., a passage of English text, or a long
> mathematical derivation. We shall consider
> such a sequence of symbols to be "simple"
> and have high a priori probability, if
> there exists a very brief description of
> this sequence—using, of course, some sort
> of stipulated description method. More

exactly, if we use only the symbols 0 and 1
to express our description, we will assign
the probability 2^{-n} to a sequence of symbols,
if its shortest possible binary description
contains n digits.

The next paragraph already makes clear that what
he will mean by a short "description" of a string
x: a program of a general-purpose computer that
outputs x.

The combination of these three ingredients:
simplicity, apriori probability, universal computer
turned out to have explosive power, forming the
start of a theory that is far from having exhausted
its potential now, 50 years later. This was greatly
helped by Kolmogorov's independent discovery that
related them explicitly to two additional classical
concepts of science: *randomness* and *information.*

There is another classical principle of assigning
apriori probabilities that has been given a new
interpretation by Solomonoff's approach: *Laplace's*
principle of indifference. This says that in the
absence of any information allowing to prefer one
alternative to another, all alternatives should be
assigned the same probability. This principle has
often been criticized, and it is indeed not easy
to delineate its reasonable range of applicability,
beyond the cases of obvious symmetry. Now in
Solomonoff's theory, Laplace's principle can be seen
revived in the following sense: if an outcome has
several possible formal descriptions (interpreted
by the universal monotonic machine), then *all*
descriptions of the same length are assigned the
same probability.

The rest of the report [14] has a groping, gradual
nature as it is trying to find the appropriate
formula for apriori probability based on simplicity
of descriptions.

The problems it deals with are quite technical in nature, that is it is (even) less easy to justify the choices made for their solution on a philosophical basis. As a matter of fact, Solomonoff later uses (normalized versions of) (2) instead of the formulas of these early papers. Here are the problems:

1) Machine dependence. This is the objection most successfully handled in the paper.

2) If we assign weight 2^{-n} to binary-strings of length n then the sum of the weights of all binary strings is infinite. The problem is dealt with in an ad-hoc manner in the report, by assigning a factor $(1 - \epsilon)^k$ to strings of length k. Later papers, in particular Solomonoff's first published paper [16] on the subject, solve it more satisfactorily by using some version of definition (1): on monotone machines, the convergence problem disappears.

3) We should be able to get arbitrary conditional probabilities in our Bayesian inference, but probability based on shortest description leads to probabilities that are powers of two. Formula (2) solves this problem as simply as it solved the previous one, but the first publication [16] did not abandon the ad-hoc approach of the technical report yet either, summing up probabilities for all continuations of a certain length (and taking the limit).

4) There are principles of induction suggesting that not only minimal descriptions (explanations) should be considered. Formula (2) incorporates all descriptions in a natural manner. Again, the ad-hoc approach, extending the sum over all descriptions (weighted as above), still is also offered in [16].

It remained for later researchers (Kolmogorov, Levin) to discover that—in certain models (though not on monotonic computers) even to within an additive constant—asymptotically, the logarithm of the apriori probability obtained this way is the same as the length of the shortest description. Thus, a rule that bases prediction on shortest explanations is not too different from a rule using the prediction fitting "most" explanations. In terms of the monotone machines, this relation can be stated as follows. For a string x, let $Km(x)$ be the length of the shortest binary string that causes the fixed optimal monotonic machine to output some continuation of x. Then

$$Km(x) - 2\log Km(x) \leq -\log M(x) \leq Km(x) \qquad (3)$$

The paper [16] offers yet another definition of apriori probability, based on a combination of all possible computable conditional probabilities. The suggestion is tentative and overly complex, but its idea has been vindicated by Levin's theorem, in [27], showing that the distribution $M(x)$ dominates all other "lower semicomputable semimeasures" on the set of infinite sequences. (Levin did not invent the universal semimeasure $M(x)$ as response to Solomonoff's work, but rather as a natural technical framework for treating the properties of complexity and randomness.) Here, the *semimeasure* property requires, for all x, the inequalities $M(x) \geq \sum_{b \in \Sigma} M(xb)$, while $M(\wedge) \leq 1$ for the empty string $\wedge$. Lower semicomputability requires that $M(x)$ is the limit of an increasing wequence of functions that is computable in a uniform way. A computable measure is certainly also a lower semicomputable semimeasure. The dominance property distinguishes Solomonoff's apriori probability among all lower semicomputable semimeasures. Levin's observation is crucial for all later theorems proved about apriori probability; Solomonoff made important use of it later.

The paper [17] considers some simple applications of the prediction formulas, for the case when the sequence to be predicted is coming from tossing a (possibly biased) coin, and when it is coming from a stochastic context-free grammar. There are some computations, but no rigorous results.

5. The Prediction Theorem

Solomonoff wrote an important paper [18] that is completely traditional in the sense of having a non-trivial theorem with a proof. The result serves as a justification of the prediction formula (2). What kind of justifications are possible here? Clearly, not all sequences can be predicted successfully, no matter what method is suggested. The two possibilities are:

1) Restrict the kind of sources from which the sequences may be coming, to a still sufficiently wide class.
2) Show that in an appropriate sense, your method is (nearly) as good as any other method, in some wide class of methods.

There is a wealth of research on inference methods considering a combination of bothe kinds of restriction simultaneously, showing typically that for example if a sequence is generated by methods restricted to a certain complexity class then a successful prediction method cannot be restricted to the same class.

Solomonoff's theorem restricts consideration to sources $x_1 x_2 \ldots$ with some computable probability distribution P. Over a finite alphabet Σ, let $P(x)$ denote the probability of the set of all infinite sequences starting with x, further for a letter b of the alphabet denote $P(b \mid x) = P(xb)/P(x)$. The theorem says that the formula $M(b \mid x_1 \ldots x_n)$, gets closer and closer to the conditional probability $P(b \mid x_1 \ldots x_n)$ as n grows—closer for example in a mean square sense (and then also with P-probability

1). This is better than any classical predictive strategy can do. More explicitly, the value

$$S_n = \sum_{x:|x|=n-1}\sum_{b\in\Sigma} p(x)(M(b\,|\,x)-P(b\,|\,x))^2$$

is the expected error of the squared probability of the *n*th prediction if we use the universal *M* instead of the unknown *P*. Solomonoff showed $\sum_{n=1}^{\infty} S_n < \infty$. (Thebound is essentially the complexity *K(P)*, of *P*, so it is relatively small for simple distributions *P*. There is no bound when *P* is not even computable.) Hence the expected squared error can be said to degrade faster then 1/*n* (provided the expectation is "smooth").

The set of *all* computable distributions is very wide. Consider for example a sequence $x_1 x_2$ ...whose even-numbered binary digits are those of π, while its odd-numbered digits are random. Solomonoff's formula will converge to 1/2 on the odd-numbered digits. On the even-numbered digits, it wil get closer and closer to 1 if *b* equals the corresponding digit of π, and to 0 if it does not. By now, several alternative theorems, and amplifications on this convergence property have appeared: see for example [7,5].

The proof relies just on the fact that *M(x)* dominates all computable measures (even all lower semicomputable semimeasures). It generalizes therefore to any family of measures that has a dominating measure—in particular, to any countable family of measures.

Despite the attractiveness of the formula, its incorporation of such a number of classical principles, and the nice form of the theorem, it is still susceptible to a justified criticism: the formula is in a different category from the sources that it predicts: those sources are computable, while the formula is not ($M(xy)/M(x)$ is the ratio of two lower semicomputable functions). But as mentioned above, the restrictions on the source and on the predictor cannot be expected to be the same, and at least Solomonoff's formula is brimming with philosophical significance.

The topic has spawned an elaborate theory of prediction in both static and reactive unknown environments, based on universal distributions with arbitrary loss bounds (rather than just the logarithmic loss) using extensions and variations of the proof method, inspiring information theorists such as Thomas M. Cover [4]. An example is the book by Marcus Hutter [5]. A related direction on prediction and Kolmogorov complexity, using various loss bounds, going by the name of "predictive complexity", in a time-limited setting, was introduced by Vladimir G. Vovk (see [26] and later works).

We noted that Solomonoff normalized his universal apriori distributions, in order to turn them into regular probability distributions. These normalizations make the theory less elegant since they take away the lower semicomputability property: however, Solomonoff never gave them up. And there is indeed no strong argument for the semicomputability of $M(x)$ in the context of prediction. In about 1992, Robert M. Solovay proved that every normalization of the universal a priori semimeasure to a measure would change the relative probabilities of extensions by more than a constant (even incomputably large) factor. In a recent paper with a clever and appealing proof, Solomonoff [25] proved that if we predict a computable measure with a the universal a priori semimeasure normalized according to his prescription, then the bad changes a la Solovay happen only with expectation going fast to 0 with growing length of the predicted sequence.

6. Universal Search

It was not until 1978, that Ray Solomonoff started to pay attention to the emerging field of computational complexity theory. In that year, Leonid Levin arrived in Boston, and they became friends. Levin had discovered NP problems around 1970, independently from Stephen Cook, and had shown the completeness of a small number of NP problems (independently of Richard Karp). For our present purpose, an NP problem is best viewed as a *search problem*. It is defined with the help of a *verification predicate* $V(x, w)$, where x is the *instance*, w is a potential *witness*, and $V(x, w)$ is true if and only if the witness is accepted. We can assume that $V(x, w)$ is computable in time linear in the size $|x|$ of the instance x (in an appropriate computation model, see later). The problem is to decide for a given instance x whether there is any witness w, and if yes, to find one. As an example, consider the problem of finding a description of length l that computes a given string x within time t on some fixed machine U. Let $x = U^t(p)$ mean that machine U computes x in time t from program p. The instance of the problem could be the string $0^l 10^t 1x$, and the verifier $V(0^l 10^t 1x, p)$ would just check whether $|p| \leq l$ and $U^t(3) = x$.

Levin's paper [6] announces also a theorem that has no counterpart in the works of Cook and Karp: the existence of an algorithm that finds a witness to an NP-complete problem in time optimal to within a multiplicative constant. Theoretically, this result is quite interesting: from then on, one could say that the question has not been *how* to solve any NP problem efficiently, only *what* is the complexity of Levin's algorithm. If there is a theorem that it works in time $g(|x|)$, then of course also the problem of whether there is any witness at all becomes decidable in time $g(|x|)$.

Levin's paper gave no proof for this theorem (a proof can be found now, for example, in [7]). There is a natural, approximate idea of the proof, though. What is special about an NP problem is that once a potential witness is guessed, it is always possible to check it efficiently. Therefore it does not harm much (theoretically, that is as long as we are willing to tolerate multiplicative constants) a good solution algorithm $A(x)$ if we mix it with some other ones that just make wild guesses. Let ρ_1, ρ_2,... be any computable sequence of positive numbers with $\sum_i \rho_i \leq 1$. We could list *all* possible algorithms A_1, A_2,..., in some order, and run them *simultaneously*, making a step of algorithm A_i in a fraction ρ_i of the time. At any time, if some algorithm A_i proposes a witness we check it. In this way, if any algorithm A_i finds witnesses in time $g(|x|)$ then the universal algorithm finds it in time $\rho_i^{-1} g(|x|)$: this is what is meant by optimality within a multiplicative constant.

In order to actually achieve the multiplicative constant in his theorem, Levin indicated that the machine model U has to be of a "random access" type: more precisely, of a type introduced by Kolmogorov and Uspensky and related to the "pointer machine" of Schönhage. He also introduced a variant of description complexity $Kt(w) = \min_{t,\ z:U'(z)\ =\ w} |z| +$ $\log t$ in which a penalty of size $\log t$ is built in for the running time t of the program z outputting the sequence w on the universal machine U. A more careful implementation of Levin's algorithm (like the one given later by Solomonoff) tries the candidate witnesses w essentially as ordered by their complexity $Kt(w)$.

Up to now, Levin's optimal algorithm has not received much attention in the computational complexity literature. In its present form, it does not seem practical, since the multiplicative constant p_z^{-1} is exponential in the length of the program z. (For time bounds provable in a reasonable sense, Hutter reduced the multiplicative constant to 5, but with a tremendous additive constant [7]. His optimal algorithm depends on the formal system in which the upper bounds are proved.) But Solomonoff appreciated it greatly, since in computing approximations to his apriori probability, this seems still the best that is available. He gave detailed implementations of the optimal search (giving probably the first written proof of Levin's theorem), in its application to computing algorithmic probability [19, 21]. These did not result in new theorems, but then Solomonoff had always been more interested in practical learning algorithms. In later projects (for example [22]) aimed at practical prediction, he defines as the *conceptual jump size* CJS of the program z the quantity t_z/p_z, where p_z is some approximation to the apriori probability of z, and t_z is its running time. The logarithm of the conceptual jump size and Levin's $Kt(w)$ are clearly related.

7. Training Sequences

Solomonoff continued to believe in the existence of a learning algorithm that one should find. He considered the approach used for example in practical speech recognition misguided: the algorithm there may have as many as 2000 tunable real number parameters. In the 1990s, he started a company to predict stock performance on a scientific basis provided by his theories. Eventually, he dropped the venture claiming that "convergence was not fast enough."

In a number of reports: [13, 15, 20, 22, 9, 24], universal search as described above is only a starting point for an array of approaches, that did not lead to new theorems, but were no less dear to Ray's heart for that. What we called "program" above can alternatively be called a "problem solving technique", or a "concept". This part of Ray's work was central for him; but the authors of the present article are closer to mathematics than to the experimental culture of artificial intelligence, therefore the evaluation poses challenges for them. We hope that the AI community will perform a less superficial review of this part of the oevre than what we can offer here.

Learning proceeds in stages, where each stage includes universal search. The conceptual jump size CJS introduced above (see [9]) continues to play a central role. Now, "probability" is used in the sense of the probability assigned by the best probabilistic model we can find in the available time for the given data. There is also an update process introducing more and more complex concepts. The concepts found useful on one stage are promoted to the status of primitives of a new language for the next stage, allowing to form more complex composite concepts (and goals). They are combined in various ways, assigning preliminarily just product probability to the composite concept. If a composite concept proves applicable with a probability beyond this initial value, it will be turned it into a new building block (with a corresponding larger probability). In this way, one hopes to alleviate the problem of excessively large multiplicative constants of universal search (see [21]).

Ray did not limit inductive inference to a model where a learner is presented a stream of experimental results. He realized that in practice, a lot of learning happens in a much more controlled situation, where there is a "teacher" (or several). Now, *supervised learning* is a well-studied set of models: in this, a teacher provides answers to some set of questions that the learner can ask. In Solomonoff's model, the teacher also *orders* the questions in increasing conceptual jump size, facilitating thereby the above concept-building process. Already the report [13] sketches a system designed to recognize more and more complex patterns, as it is being fed a sequence of examples of gradually increasing complexity.[1] Ray spent many years working out some examples in which a learning algorithm interacts with a training sequence. The examples were of the type of learning a simple language, mainly the language of arithmetic expressions. By now, there are systems in AI experimenting with learning based on universal optimal search: see Schmidhuber in [12] and other works.

We are not aware of any *theoretical* study that distinguishes the kind of knowledge that the teacher can transmit directly from the one that the student must relearn individually, and for which the teacher can only guide: order problems by complexity, and check the student answers. The teacher may indeed be in conscious possession of a network of concepts and algorithms, along with estimates of their "conceptual jump size", and we should assume that she communicates to the student directly everything she can. (The arithmetic algorithms, Ray's main example, can certainly be fed into a machine without need for learning.) But it appears that in typical realistic learning, the directly, symbolically transferable material is only a very incomplete projection of the mental models that every pupil needs to build for himself.

---

[1] Marvin Minsky considers that the practical potential of the pattern recognition algorithms in this work of Ray still has not received the attention it deserves.

References

[1] Rudolf Carnap. *Logical Foundations of Probability*. University of Chicago Press, 1950.

[2] Gregory J. Chaitin. On the length of programs for computing binary sequences, II. *Journal of the ACM*, 16: 145–159, 1969.

[3] Noam Chomsky. Three models for the description of language. *IRE Trans. Inform. Theory*, 2(3): 113–124, September 1956.

[4] Thomas M. Cover. Universal gambling schemes and the complexity measures of Kolmogorov and Chaitin. In J.K. Skwirzynski, editor, *The Impact of Processing Techniques on Communication*, pp. 23–33. Martinus Nijhoff, 1985. Stanford University Statistics Department Technical Report # 12, 1974.

[5] Marcus Hutter. *Universal Artificial Intelligence: Sequential Decisions Based on Algorithmic Probability*. Springer-Verlag, Berlin, 2005.

[6] Leonid A. Levin. Universal sequential search *problems. Problems of Inform. Transm.*, 9(3): 255–256, 1973.

[7] Ming Li and Paul M.B. Vitányi. *Introduction to Kolmogorov Complexity and its Applications (Third edition)*. Springer Verlag, New York, 2008.

[8] Marvin L. Minsky. Problems of formulation for artificial intelligence. In R.E. Bellman, editor, *Proceedings of the Fourteenth Symposium in Applied Mathematics,* pp. 35–45, New York, 1962. American Mathematical Society.

[9] Wolfgang Paul and Raymond J. Solomonoff. Autonomous theory building systems. In P. Bock, M. Loew, and M. Richter, editors, *Neural Networks and Adaptive Learning*, pp. 1–13, Schloss Reisenburg, 1990.

[10] Jorma J. Rissanen. A universal prior for integers and estimation by minimal description length. *Annals of Statistics*, 11(2): 416–431,1983.

[11] Jorma J. Rissanen. *Stochastic Complexity in Statistical Inquiry*. World Scientific, London, U.K., 1989.

[12] Jürgen Schmidhuber. Optimal ordered problem solver. *Machine Learning*, 54: 211–254, 2004.

[13] Raymond J. Solomonoff. An inductive inference machine. In *IRE Convention Record, Section on Information Theory,*

pp.56–62, New York, 1957. Author's institution: Technical Research Group, New York 3, N.Y.

[14] Raymond J. Solomonoff. A preliminary report on a general theory of inductive inference. Technical report ZTB-138, Zator Company, Cambridge, MA, 1960.

[15] Raymond J. Solomonoff. Training sequences for mechanized induction. In M. Yovits, editor, *Self-organizing systems*, 1961.

[16] Raymond J. Solomonoff. A formal theory of inductive inference I. *Information and Control*, 7: 1–22, 1964.

[17] Raymond J. Solomonoff. A formal theory of inductive inference II. *Information and Control*, 7: 225–254, 1964.

[18] Raymond J. Solomonoff. Complexity-based induction systems: Comparisons and convergence theorems. *IEEE Transactions on Information Theory*, IT-24(4): 422–432, July 1978.

[19] Raymond J. Solomonoff. Optimum sequential search. Technical report, Oxbridge Research, Cambridge, MA,1984.

[20] Raymond J. Solomonoff. Perfect training sequences and the costs of corruption—a progress report on inductive inference research. Technical report, Oxbridge Research, Cambridge, MA, 1984.

[21] Raymond J. Solomonoff. The application of algorithmic probability to problems in artificial intelligence. In L.N. Kanal and J.F. Lemmer, editors, *Uncertainty in Artificial Intelligence*, Advances in Cognitive Science, AAAS Selected Symposia, pp. 473–491, North-Holland, 1986. Elsevier.

[22] Raymond J. Solomonoff. A system for incremental learning based on algorithmic probability. In *Proceedings of the Sixth Israeli Conference on Artificial Intelligence, Computer Vision and Pattern Recognition,* pp. 515–527, Tel Aviv, 1989.

[23] Raymond J. Solomonoff. The discovery of algorithmic probability. *Journal of Computer System sciences*, 55(1): 73–88, 1997.

[24] Raymond J. Solomonoff. Progress in incremental machine learning. Technical Report 03-16, IDSIA, Lugano, Switzerland, 2003. Revision 2.0. Given at NIPS Workshop on Universal Learning Algorithms and Optimal Search, Dec. 14, 2002, Whistler, B.C.,Canada.

[25] Raymond J. Solomonoff. The probability of "undefined" (non-converging) output in generating the universal probability distribution. *Information Processing Letters*, 106(6): 238–246, 2008.

[26] Vladimir G. Vovk. Prediction of stochastic sequences. *Problems of Information Transmission*, 25: 285–296, 1989.

[27] Alexander K. Zvonkin and Leonid A. Levin. The complexity of finite objects and the development of the concepts of information and randomness by means of the theory of algorithms. *Russian Math. Surveys*, 25(6): 83–124, 1970.

Index

About the Author

Dr. Bradley S. Tice, FRSS

Dr. Tice has professional interests in the areas of statistical physics that relate to telecommunications and in the study of natural languages. Dr. Tice is a Fellow of The Royal Statistical Society.

Images

Image 1 Enigma Machine

Image 2 Enigma Machine

Image 3 Enigma Machine

Image 4 Difference Engine Number 2

Image 5 Enigma Machine

For Product Safety Concerns and Information please contact our EU
representative GPSR@taylorandfrancis.com
Taylor & Francis Verlag GmbH, Kaufingerstraße 24, 80331 München, Germany

www.ingramcontent.com/pod-product-compliance
Ingram Content Group UK Ltd.
Pitfield, Milton Keynes, MK11 3LW, UK
UKHW021123180425
457613UK00005B/195